# Glossary of Cereal Science and Technology

**First Edition**

Other books by the same author—

Chemistry and Technology of Cereals as Food and Feed
First and Second Editions

Glossary of Milling and Baking Terms

Bakery Technology and Engineering
First and Second Editions

Equipment for Bakers

Cereal Technology

Cereal Science

Food Texture

Water in Foods

Bakery Technology

Ingredients for Bakers

Cookie and Cracker Technology
First, Second, and Third Editions

Formulas and Processes for Bakers

Technology of the Materials of Baking

Snack Food Technology
First, Second, and Third Editions; Japanese Edition

\* \* \* \* \* \*

This book was designed and manufactured to give you maximum value and long use. It is printed on good quality, heavy base-weight paper of superior opacity and whiteness, so that it will last longer and be easier to read. The type face is New Century Schoolbook, an open style which is esthetically pleasing and can be scanned quickly and accurately. The book has been bound by a process called "Smyth sewn in signatures," a method selected to give a sturdy book with convenient handling properties, and the binding was made of stronger-than-usual papers and heavier boards to support the pages without sagging. The cover is plastic impregnated cloth that is resistant to wear, stains, and other damage; its gold stamping helps to make the book an attractive addition to your library.

GLOSSARY OF CEREAL SCIENCE AND TECHNOLOGY

by

SAMUEL A. MATZ, PH. D.

*President, Pan-Tech International, Inc. Formerly, Vice President for Research, Development, and Compliance, Ovaltine Products, Inc. At one time, Vice President for Research and Development, Robert A. Johnston Co.; Technical Director of the Refrigerated Dough Program, Borden Foods Co.; Chief of the Cereal and General Products Branch, Quartermaster Food and Container Institute for the Armed Forces; Chief Chemist, Harvest Queen Mill and Elevator Co.; Instructor, Department of Flour and Feed Milling Industries, Kansas State University; Chemist, Iglehart Mills Division of General Foods.*

PAN-TECH INTERNATIONAL, INC.
P. O. BOX 4548
MC ALLEN, TEXAS 78502
1995

Copyright 1995 by SAMUEL A. MATZ

ISBN 0-942849-25-6

All rights reserved. No part of this work covered by the copyright hereon may be reproduced or used in any form or by any means — graphic, electronic, or mechanical, including photocopying, recording, taping, or information storage and retrieval systems — without written permission of the copyright owner.

Library of Congress Cataloging-in-Publication Data

Matz, Samuel A.
    Glossary of cereal science and technology / by Samuel A. Matz --
1st ed.
    p.   cm.

ISBN 0-942849-25-6
    1. Cereal products--Dictionaries.   2. Cereal products industry--Dictionaries. 3. Grain--Milling--Dictionaries. I. Title. TP434.M38   1995
664'.7'03--dc20                                                       94-48585
                                                                        CIP

PRINTED IN THE UNITED STATES OF AMERICA

## INTRODUCTION

According to the unabridged third edition of Merriam-Webster's "Webster's New International Dictionary," a **Glossary** is, "A collection of glosses, or explanations of words and passages of a work or author; a partial dictionary of words and passages of a work, an author, a dialect, art or science, explaining technical terms or uncommon words." The book you are reading is not intended to be an all-purpose or all-inclusive dictionary. Definitions are restricted to uses of the respective word, phrase, or acronym in the many aspects of cereal science and technology. As an example of this positioning, we find the word "blind" is herein defined in three distinct ways, none of them being "sightless" (or its equivalent).

This Glossary contains definitions of words and phrases that may be encountered in publications on agronomy, dry-milling, wet-milling, brewing, baking, macaroni production, breakfast cereal and snack manufacture, dietetic food formulation, and allied technologies and industries. Further, the definitions are specifically applicable to the texts of the several books in these fields that I have written. Words and phrases describing processes, machinery, utensils, quality tests, ingredients, and products are among the terms that were selected for inclusion.

The words defined in the following pages have placed in alphabetical order, with spaces, hyphens, apostrophes, accent markings, and capitalization ignored in their positioning. This is not in accord with the practices in some dictionaries and indexes, but it seemed to me to be the most helpful alternative for the present purpose and, for that matter, the most logical method for alphabetically organizing most lists of words.

Simplifications of foreign alphabets have been necessary in order to make the glossary more convenient for readers accustomed to the English alphabet. Letters found only in non-English alphabets are taken to be the nearest equivalent in the English alphabet and are positioned accordingly; thus "ægis" would fall somewhere between "adulterated" and "aflatoxin," while "Ångstrom" would be placed after "amortize" and before "apple." German nouns are not capitalized. Transliteration of words from other alphabet versions may vary from that found elsewhere. The treatment of words originating in China or in other Asian countries is of necessity a compromise among several possible choices.

The extensive collection of terms defined in this glossary includes numerous foreign language entries. The principal geographic area of use (if other than the USA) or of the term's origin is denoted by a country name, printed in italics immediately following the defined word. The use of geographic areas rather than specific language groups, i.e., *France* instead of *French*, was deliberate, though it may be more vulnerable to the complaint that it lacks precision. The simplest way of justifying the procedure is to point out the many differences in meanings of certain words used in the UK and the USA (e.g., "biscuit"), even though "English" is acknowledged to be the dominant language in both countries. This practice did, however, necessitate some compromises and omissions — for example, the names of many cereal products are different in Mexico and Spain (see "tortilla"); I am far more familiar with Mexican than Spanish usage, so there may have been some bias in selecting the former area when attributions were assigned. If a word or phrase has become widely accepted in USA scientific, industrial, or marketing circles, no indication of linguistic origin has been given. After all, this is not intended to be an etymological treatise.

All dictionaries, glossaries, and other collections of definitions draw extensively from preceding works of the same type, and this volume is no exception. The references consulted by the author were too numerous to be listed here, but the greatest reliance was placed on my own books.

Samuel A. Matz<br>
McAllen, Texas<br>
October 1, 1995

## -A-

**Abernethy biscuit** — *UK* a kind of small round cookie or cracker containing caraway seeds; seldom seen nowadays.

**abrasive milling** — reducing a grain to meal or flour, or removing its external layers, by contacting the kernel with abrasive (sandpaper-like) surfaces. Applied particularly to rice and barley.

**absolute** — (noun) the highly concentrated extract of a botanical (often a flower) that has been treated to remove any matter insoluble in alcohol (e.g., waxes). May be liquid, semi-solid, or (less often) solid.

**absorption** — (1) The amount of water required to be added to a particular flour in order for it to function optimally in some application (usually the amount required to make the best possible bread dough), expressed as a percent of the flour weight. (2) Amount of oil or fat retained by a food product that has been fried.

**accelerated rancidity test** — any method of estimating the storage life of fats or fat-containing products by speeding up the onset and progress of rancidity in a sample of the material. Usually involves increasing the temperature, oxygen tension, light intensity, etc.

**aceite** — *Spain* oil.

**acesulfame K** — an artificial sweetener, about 200 times as sweet as sucrose.

**acetic acid** — an organic acid, $CH_3COOH$. The pure substance (glacial acetic acid) is a colorless, pungent liquid congealing at cool room temperatures. Acetic acid of commerce is an aqueous solution containing various percentages of the compound. Vinegar contains 4.5 to 12% (typically 5%) acetic acid; it has been used as a preventative of rope spoilage in bread.

**acetylated monoglycerides** — a class of chemicals, some of which have emulsifying effects when added to bread doughs.

**acid** — variously defined in different times and schools, but, in the broadest chemical sense, any substance capable of donating protons to bases. For the technologist, a more useful definition is: a chemical compound yielding hydrogen ions when in water solution.

**acid conversion method** — a hydrolysis process for breaking down corn starch into glucose and other saccharides in the form of corn syrup.

**acid creams** — substances or mixtures of substances that can react with sodium bicarbonate to form satisfactory leavening systems in doughs and batters. Usually, the active substance is mixed with diluents, stabilizers, etc., to form proprietary mixtures that are superior to cream of tartar in some respects.

**acidic** — (1) Having a pH of less than 7. (2) Having a tart or sour taste.

**acid potassium tartrate** — cream of tartar.

**acidulant** — a substance added to a system such as a beverage in order to increase the acid taste and/or reaction of the system.

**Acid Value** — a measure of the amount of free fatty acids in a sample of fatty material, reported as the number of milligrams of potassium hydroxide required to neutralize the free fatty acids in a gram of fat.

**acid whey** — a form of dried whey that has a distinctly acidic reaction resulting from lactic acid or other acidic material present in the cheese vat from which the whey is taken. In general, less desirable than sweet whey for breadmaking purposes.

**acqua** — *Italy* water.

**acrid** — harsh or bitterly pungent in taste or aroma.

**activated charcoal** — charcoal that has been prepared under conditions, or further treated, so as to give a highly adsorbent powder or granules. Much used in water purification to remove taste, color, and odor promotants.

**Activated Dough Development** — according to the Flour Milling and Baking Research Association of the UK, a process of breadmaking in which no time has been allocated for bulk fermentation; adequate development of the dough is achieved by adding fairly large amounts of cysteine, ascorbic acid, and potassium bromate.

**active dry yeast** — a preparation of live bakers' yeast that has been dried to about 6% to 8% moisture; usually offered in the form of granules or small pellets. Preservatives and other additives may be present.

**Active Oxygen Method** — an accelerated rancidity test in which a sample of fat is held at 98°C while air is bubbled through it at a specified rate. The endpoint is reported as hours needed to reach a peroxide value of 100 meq/Kg. There is a variable relationship between AOM hours of an oil and the actual shelf-life of a product containing the oil.

**açúcar** — *Portugal* sugar.

**additive** — according to the FDA: any substance, the intended use of which results or may reasonably be expected to result, directly or indirectly, in its becoming a component of, or otherwise affecting, the characteristics of any food.

**adjunct** — something joined or added to another thing, but not essentially a part of it. Frosting, filling, decorations, and the like are adjuncts to a cake, for examples.

**adornar** — *Spain* (v.) to decorate, as in adornar el pastel (to decorate a cake).

**adulterate** — to clandestinely add some foreign or inferior substance to a food or ingredient, usually for the purpose of making the food material cheaper to produce, as added water would be an adulterant in milk.

**æble** — *Denmark* apple.

**æblekage** — *Denmark* cake made with layers of buttered pumpernickel crumbs, sherry-soaked macaroons, and flavored applesauce. Topped with whipped cream.

**æbleskive** — *Denmark* apple fritter, usually served with jam.
**æg** — *Denmark* egg.
**æggeblomme** — *Denmark* egg yolk.
**aeration** — the leavening of a dough or (more often) a batter by mixing in air or (less often) by injecting compressed gas. Sometimes applied to any kind of leavening action.
**aerobic** — applied to microorganisms that require uncombined oxygen ($O_2$) for growth; contrasted with anaerobic organisms that grow best in atmospheres of very low oxygen content.
**aflatoxin** — one of a related group of organic compounds produced by certain species of *Aspergillus*; some of these compounds are extremely poisonous and/or carcinogenic.
**agar** — (agar agar) powder derived from sea plant extractives; forms a very firm gel with large percentages of water. Used in decorating jellies, etc.
**agemnono** — *Japan* deep-fat frying.
**ägg** — *Sweden* egg.
**ägghvitor** — *Sweden* egg whites.
**agglomeration** — a process whereby several small particles are adhered together to form a larger particle; wheat flour can be agglomerated to produce an ingredient less subject to clumping when water is added.
**äggular** — *Sweden* egg yolks.
**aging** — a step in the milling process in which flour is stored for a considerable time after grinding so that its original creamy tint is greatly reduced by natural reactions and its baking quality is improved. Seldom, if ever, used in modern mills in the U.S.A.
**agitator** — the part (paddle, whip, dough hook, etc.) of a mixer that causes turbulence in the bowl's contents.
**agnolotti** — *Italy* a sort of semi-circular ravioli, traditionally stuffed with chopped meat and vegetables.
**agua** — *Spain* water. *Portugal* "água."
**air-belt purifier** — a purifier (q.v.) in which the same air is recirculated continuously through the sieve, fan, and dust collector.
**air classification** — a process by which the particles in a mill stream are separated according to size and density by cyclone separators instead of sieves.
**air infiltration** — the unwanted entry of non-conditioned air into a space that is being held at constant temperature or humidity (or both).
**air leavening** — increase in volume of a dough or batter due to air bubbles that have been whipped or beaten into the dough or batter or into a component such as meringue.
**ajo** — *Spain* garlic.
**akhroot** — *India* walnut.
**alatizmeno** — *Greece* salted.

**Alberger salt** — a small-flake salt that has been made from brine treated to remove a large part of its calcium and magnesium ions.
**albumen** — egg white.
**albumin** — a type of protein, characterized by its solubility and chemical properties.
**albumin rest** — or, "protein rest." In brewing, a relatively quiescent period when the mash is held at about 113°F to facilitate proteinase action.
**alcohol** — in general, an organic chemical compound containing the hydroxyl group -OH. Specifically, ethanol (grain alcohol).
**al dente** — describes the texture of pasta that has been cooked just long enough to eliminate the raw core, but still leave a detectable chewiness.
**ale** — an alcoholic beverage similar to beer, but generally made using a top-fermenting yeast and often containing more ethanol than beer. There are many variations.
**aletria** — *Portugal* thin noodles, vermicelli.
**alemares** — *Mexico* pretzel shapes made from pan fino dough and sprinkled with coarse sugar before baking.
**alginates** — salts of alginic acid found in certain kinds of algae and used as viscosity improvers in foods and beverages.
**alimentary paste** — pasta; macaroni/spaghetti products of any shape or size, i.e., extruded, sheeted and cut or otherwise formed shapes of dough made from water and durum semolina or wheat farina. Thus, egg noodles and ravioli would not be alimentary pastes while dough strips in the shape of noodles but containing no eggs or other adjuncts would be called alimentary paste. In a wider sense, and probably the most common usage in the U.S., any shape made principally from a granular wheat or durum raw material (even though it contains adjuncts) and used in the same way as macaroni, spaghetti, noodles, etc.
**alimento** — *Spain* food.
**alitame** — a dipeptide-based amide about 2,000 times sweeter than sugar.
**alkaline water retention capacity** — a test that has been recommended for assessing the baking quality of a sample of wheat flour, meal, or grain.
**allspice** — the dried, nearly ripe fruit of *Pimenta dioica*, appearing as a reddish-brown, pea-sized seed with a fairly smooth surface. Has a strong clove note with nuances of cinnamon and nutmeg. In the spice trade, it is often called "pimento," causing confusion with "pimiento," a type of bell pepper.
**allumettes** — *France* rectangles of puff pastry filled with a sweet or savory mixture.
**Alma tea cake** — *UK* a kind of cookie cooked on a griddle.
**almendra** — *Spain* almond.
**almibar** — *Spain* sirup.
**almond** — a nut much favored for use in and on bakery products, the kernel or seed of a small tree. Available in many sizes and varieties.

**almond butter** — finely ground, roasted almonds.
**almond paste** — almonds ground with sugar to yield a stiff paste; sometimes a binding ingredient is added. Used as an ingredient in bakery fillings, etc.
**aloo** — *India* potato.
**alpha-amylase** — an enzyme having as its major useful function the splitting of starch molecules at random points, forming smaller molecules of widely varying size, e.g, the so-called dextrinizing enzyme of malt.
**altamura bread** — *Italy* a sourdough bread made from mixtures of durum and hard wheat flour.
**alum** — several aluminum sulfate salts have been called alum, but for the baker it nearly always means sodium aluminum sulfate, which is used as the acid reacting component of some baking powders.
**aluminum** — a light relatively soft metal of silvery color. It is a good conductor of heat and electricity. When exposed to air and moisture, it quickly becomes covered with a thin hard coating of oxide that resists further corrosion.
**alveograph** — instrument for measuring the extensibility of dough; provides an indication of baking quality. As a standard disc of dough is blown into a bubble, pressure change and bursting pressure are charted vs. time.
**amand** — *France* almond
**amandel** — *Netherlands* almond.
**amandel broodje** — *Netherlands* sweet roll with almond-paste filling.
**amaranth** — a herbaceous plant that produces small edible seeds on a sorghum-like head; it is a grain but not a cereal. The U.S. crop is quite small, but the grain is an important food in some Asian and African countries. Has been used as a "grain" ingredient in multigrain breads. The young, tender leaves are also eaten, being cooked like spinach.
**amaretti** — *Italy* cookies of the macaroon type made with bitter almonds and having a porous structure; baked dry and crisp. Amarettini are smaller cookies of the same general type.
**amasar** — (v.) *Spain* to knead, as in "amasar la pasta," to knead the dough.
**ambient** — surrounding or encompassing, as ambient conditions.
**ameixa seca** — *Portugal* prunes.
**amêndoa** — *Portugal* almond.
**amendoim** — *Portugal* peanut.
**Amflow Process** — a trade name used by the AMF Company for their version of continuous breadmaking plants.
**amighdhalo** — *Greece* almond.
**amighdhaloto** — *Greece* marzipan.
**amino acids** — organic compounds containing both amino and carboxyl groups attached to a carbon chain. Proteins are made up of many such groups. Some of the amino acids are essential nutrients for humans.

**ammonia** — incorrect name for the ammonium bicarbonate-ammonium carbonate mixture sometimes used for leavening cookies, etc. Ammonia gas is never used as an ingredient.

**ammoniated glycyrrhizin** — a powerful sweetener made by reacting a licorice extract with ammonia.

**ammonium bicarbonate** — common name for a mixture of compounds that decomposes into ammonium and carbon dioxide when heated; used as a leavener in cookies and the like.

**ammonium carbamate** — a chemical compound allegedly forming a part of some commercial preparations of "ammonium bicarbonate."

**ammonium chloride** — a chemical compound sometimes used in yeast foods to provide a source of nitrogen for yeast growth.

**ammonium phosphate** — a nitrogen source for yeast growth used in dough improvers and in nutrient broths.

**amylase** — an enzyme that can hydrolyze starches.

**amylograph** — an instrument used to determine change of viscosity with time in a heated mixture of water and a starchy material such as flour.

**amylolytic enzymes** — those enzymes that hydrolyze starches and similar glucose polymers.

**amylopectin** — starch molecules made up of glucose units chemically combined in branching chains.

**amylose** — starch molecules made up of glucose units chemically combined in long unbranched chains.

**analato** — *Greece* unsalted.

**ananás** — *Spain* pineapples, in some localities this name is restricted to a smaller, sweeter type of pineapple, the others being called "piñas."

**ande** *India* egg.

**anaerobic** — indicates an organism (anaerobe) that can grow and reproduce in environments having very low oxygen tension. Obligate anaerobes are actually inhibited by significant concentrations of $O_2$.

**andruty** — *Poland* wafer cookies.

**angeer** — *India* figs.

**angoor** — *India* grapes.

**angel food cake** — a highly aerated (low density) white cake made principally from whipped egg whites, flour, and sugar. Often contains a whipping aid such as cream of tartar. Never contains shortening.

**angel food pan** — pans in which angel food cakes are baked; round with high, outward-sloping sides and a vertical cylindrical tube in the middle.

**angelica** — a herbaceous plant the leaves and stems of which are candied and used as decorations on cakes, etc. Not at all common in U.S.A. practice. Various parts of the plant have been used to flavor liqueurs.

**angle of difference** — a simple test for estimating the flowability of a powder; the difference between the angle of repose and the angle of fall.

**angu** — *Portugal* cornmeal (or cassava-root flour) boiled in water and salt.
**anydroglucose unit** — the $C_6H_{10}O_5$ unit that is bound to other such units to make up the starch molecule.
**anhydrous** — not containing moisture.
**animal crackers** — also, "menagerie biscuits." Lean formula, chemically leavened cookie doughs formed on a cutting machine into shapes resembling (to the active imagination of a child) the outlines of animals.
**anise** — the dried fruit of *Pimpinella anisum*. Has a flavor often described as "like licorice." In the bakery, this spice is used mostly in cookies, especially those common to Italian cuisines.
**anisidine value** — an indication of the extent of oxidative deterioration of fats and oils based on spectrophotometric measurements of the aldehydes present in the lipid.
**annatto** — a food coloring preparation made by extracting the coating of the seed of *Bixa orellanna* trees. Yellowish to reddish-orange, and oil soluble.
**antimicrobial agent** — an additive that deters growth of microorganisms.
**antinutritional factors** — chemicals or other factors present in certain food ingredients that can hinder the utilization of nutritional materials present in the finished food. Antitrypsin factors are examples.
**antioxidant** — a substance that will reduce the rate at which a fat or oil becomes rancid due to oxidation.
**antitrypsin factors** — chemicals found in certain grains and other food raw materials that hinder the digestive action of the enzyme trypsin; known to be present in rye, for example.
**apocarotenal** — an orange-yellow pigment related to the carotenes.
**appa** — *Sri Lanka India* also, "hoppers." Bowl-shaped rice pancakes, leavened with the fermented sap of the coconut palm (toddy).
**appel** — *Netherlands* apple.
**appel beignet** — *Netherlands* apple fritter.
**appel bol** — *Netherlands* apple dumpling.
**appel flap** — *Netherlands* puff pastry pocket containing an apple slice.
**appel gebak** — *Netherlands* apple cake.
**appelsiini** — *Finland* orange.
**apple** — the fruit of any tree of the genus *Malus*. There are hundreds of varieties, including perhaps 20 or 30 common commercial types, each of them having its particular advantages and disadvantages for baking purposes. Careful selection of type and condition of this ingredient is very important to the qyality of the finished baked product. Can be obtained fresh, frozen (with or without sweeteners), canned, dehydrated pieces (hot air- or freeze-dried), and spray-dried powders.
**apple pan dowdy** — raw sliced apples mixed with sweeteners, spices, and thickeners, covered with a relatively thick streusel-like topping of brown sugar, butter, and flour, then baked.

**applicator** — a person or machine that applies something to another thing.

**apprêt** — *France* proofing.

**apprêt sur couche** — *France* placing pieces of dough between the folds of a cloth for proofing.

**aprikoosi** — *Finland* apricot.

**aquaculture** — the commercial culture of sea, lake, and river foodstuffs such as fish, oysters, and seaweed.

**aqueous** — containing water.

**araban** — a pentosan, i.e., a polymer of arabinose units; some arabans form gummy materials in the presence of water.

**arabinoxylans** — a polysaccharide found in cell walls of the parenchymatous and lignified tissues of the wheat plant.

**arachide** — *Italy* peanuts.

**arancini** — candied orange peel in strips.

**arap pidesi** — *Turkey* a pocket bread, like pita. Yeast-leavened.

**aravositeleo** — *Greece* corn oil.

**arepa** — *Spain/Venezuela* flat bread made of masa and processed like a tortilla, except thicker.

**arme riddere** — *Norway* French toast.

**aromatic compounds** — (1) Chemically, compounds that contain at least one benzene ring in their structure. (2) Compounds that have a fragrance or smell, usually restricted to those with pleasant odors.

**aroo** — *India* peach.

**arrack** — liquor distilled from fermented palm sap, molasses, or rice mash.

**arrowroot** — a starchy preparation obtained from certain tropical plants (esp. *Maranta arundinaceae*) and used as an ingredient for thickening sauces and fillings. At one time, it was considered a kind of health food, but not a common ingredient in modern U.S.A. baked goods.

**arroz** — *Spain Portugal* rice.

**artificial** — a substance or thing produced by human agency for the purpose of imitating a natural material or object.

**artificial flavors** — a flavoring material designed to replace or supplement a flavor (or flavor note) representative of a natural food; even though a flavoring contains only natural substances, it is artificial so far as the intended use is concerned. Thus grape juice added to raspberry jam for the purpose of imparting an apparently more intense raspberry flavor would be an artificial flavor.

**asbestos** — a mineral found in fibrous form, valuable because of its resistance to flame and its low heat conductance. Now in disrepute because of alleged bad effects on health.

**ascorbic acid** — vitamin C; occurs naturally in many foods but is also made synthetically in large quantities for use as a nutritional supplement and as an improving agent in yeast doughs.

**aseptic** — describes a closed system, as a beverage in a sealed bottle, that has been rendered essentially free of microorganisms by heating, radiation, or other means.

**ash** — the inorganic material left after flour (or other organic material) is burned. Frequently included in flour specifications as a criterion of the extraction rate.

**aspartame** — a synthetic nutritional sweetener about 100 times as sweet as sucrose, sold in the form of a white, water-soluble powder.

**Aspergillus** — a genus of common fungi causing spoilage of bakery products and other moist foods; their bluish or greenish mycelia are frequently found on outdated bread.

**aspic** — a jelly, usually made from gelatin and broth, in which meat, fish, fruits, etc. are suspended. Sometimes used as a component of meat pies or fruit covered torts, etc.

**aspiration** — use of controlled velocities of directional air streams to separate particles having different resistances to air flow.

**aspirator** — a grain cleaning apparatus that utilizes the separating power of air currents to remove low density impurities such as dust, light chaff, and bran particles from grains or other granular material.

**atmospheric pressure** — force exerted on an object by the air surrounding it; at sea leavel. Decreases with increasing altitude and fluctuates with changing atmospheric conditions.

**atta** — *India* whole wheat flour.

**attorta** — *Italy* flaky pastry filled with fruit and almonds.

**Australian no-time dough** — a specific type of breadmaking process in which the need for fermentation periods is allegedly eliminated by using very intensive mixing procedures, and high levels of oxidizers and reducing agents, as well as other changes.

**autolysis** — self-digestion of cells, noticeable in compressed yeast that has been stored too long or at elevated temperatures, when the mass becomes brownish in color, semi-liquid in consistency, and unpleasant in odor.

**automation** — (1) The technique of making a process or system automatic. (2) Automatically controlled operation of an apparatus, process, or system by electronic means or otherwise.

**aveia** — *Portugal* oats.

**avela** — *Portugal* hazelnut.

**avellana** — *Spain Italy* hazelnuts, filberts.

**avgho** — *Greece* egg.

**azodicarbonamide** — a synthetic improving or maturing agent that strengthens dough by facilitating the linking of gluten molecules.

**azúcar** — *Spain* sugar.

## -B-

**baba** — a dessert cake made from rich yeast dough containing raisins, often served after it has been soaked with a sweetened rum-flavored syrup. Traditionally baked in a cylindrical mold lined with shredded almonds. Russia and other countries speaking languages of the Slavic group use similar words to describe different kinds of cakes.

**baba romovaya** — *Russia* sponge cake onto which rum or a rum-flavored syrup has been poured.

**baba tatlisi** — *Turkey* a rich baba containing raisins and apricot puree; has been described as a "Turkish savarin."

**Babcock test** — an old test much used in the dairy industry for determining the percentage of butterfat in a sample of milk or cream; involves treating the milk with acid and centrifuging it.

**babka** — *Poland* yeast or baking powder leavened cake; typical shape is taller than wide, narrower at the top; many different formulas.

**babka chlebowa** — *Poland* black-bread babka; egg-leavened rich cake containing a substantial amount of crumbs from rye bread.

**babka maslana** — *Poland* cake with high content of butter; leavened with beaten eggs.

**baby corn** — also, "miniature corn." Small, whole, husked corn on the cob, a variety of sweet corn. Used as a vegetable in some Asian cuisines. Commercially available only in cans.

**Bacillus mesentericus** — the bacteria responsible for spoilage of the rope type in bakery products. Their spores often survive the heat they encounter inside bread loaves during baking.

**backbone** — an obsolete term for the principal component of a flour blend.

**backobst** — *Germany* dried fruit.

**backpulver** — *Germany* baking powder.

**bacterial amylase** — an enzyme derived by extracting certain kinds of proteins from suitable types of bacterial cultures; differs from many other amylases in that it retains a large part of its activity even after being subjected to high temperatures.

**badaam** — or, "badem." *India* almonds.

**badem ezmesi** — *Turkey* almond paste, often flavored with rosewater.

**bademi** — *Serbia/Croatia* almonds

**bademli empare** — *Turkey* sweet almond cakes/cookies; chemically leavened; coated with a sweet syrup after baking.

**bademli kurabiye** — *Turkey* small, crescent-shaped cookies with a high content of finely ground almonds.

**badem tatlisi** — *Turkey* a light, spongy almond cake moistened with sweet syrup. High in egg content.

**baffle** — (n) a plate or other obstruction inserted in a duct or chamber to direct the flow of gases or liquids. For example, baffles are used to direct the flow of hot gases in ovens, promoting either the uniformity or non-uniformity (zones) of temperatures inside the baking chamber.

**bagel** — a doughnut-shaped (i.e., a thick annulus or ring), yeast-fermented bread product, that has a characteristic brown smooth surface resulting from its immersion in hot water shortly before it is baked; it usually has a dense, tough crumb.

**bagepulver** — *Denmark* baking powder.

**bagger** — a machine that inserts any bakery product into a plastic or paper bag; generally refers to a bread bagging machine.

**bagging out** — depositing cookie dough, icing, or other paste out of a pastry bag.

**baguette** — *France* the most common bread loaf shape in France. Typically, a long (about 2 ft.) narrow, approximately cylindrical loaf weighing between 12 and 24 oz.

**bain marie** — a double boiler used to keep mixtures hot but below boiling temperatures.

**baiser** — *France* a baked meringue.

**bajra** — *India* millet.

**bakad** — *Sweden* baked.

**bake** — cook by dry heat in an oven or similar device.

**baked Alaska** — a mound of ice cream on a layer of cake, both covered with a thick layer of meringue and baked briefly in a hot oven until the meringue is slightly browned.

**bakelse** — *Sweden* pastry, fancy cake.

**bake-off format** — a production process in which at least 95% of a bakery's output is made from frozen dough that is either purchased or made in the bakery's own plant.

**bake-out** — the amount of weight loss undergone by a dough or batter during its passage through the oven.

**bakers' cheese** — in composition, like cottage cheese without the latter's cream dressing; in form, appears white, dry, granular, and elastic. Usable as an ingredient in many fillings, often replacing ricotta.

**bakers' chocolate** — roasted and finely ground cocoa bean nibs without additives; synonymous with chocolate liquor and bitter chocolate.

**bakers' flour** — a name formerly applied to first clears.

**bakers' percent** — the weight of individual ingredients expressed as a percentage of the weight of flour in the formula. Thus, a dough made from 100 lbs flour and 60 lbs water would have 60% ingredient water (60% absorption).

**bakers' yeast** — a commercial preparation consisting mostly of living cells of *Saccharomyces cerevisiae*.

**Bakewell pudding** — *UK* a sweet pudding made of jam, almonds, and pastry or crumbs.

**Bakewell tart** — *UK* an almond cake with a filling of raspberry preserves.

**baking powder** — a mixture of sodium bicarbonate and an acid reacting substance, with other relatively inert ingredients acting as diluents and stabilizers. Used widely as a leavening system in both home and commercial baking.

**baking powder biscuit** — chemically-leavened hot bread made in individual portion size, usually as a small flat-topped round cake. The dough is lean, and the finished product has a soft white crumb and a light brown crust on the top but not on the sides. This product can be found in many types that vary in shape, size, density, and other characteristics.

**baking sheets** — flat metal sheets on which dough pieces can be baked. They are found in a wide range of sizes, sometimes with low-rise rims, and can be constructed of iron, steel, or aluminum; the term is generally not applied to pans provided with depressions for locating dough pieces.

**baking soda** — commercial sodium bicarbonate; available in several different particle sizes.

**baking test** — also, "standard baking test." A procedure for estimating the overall quality of a flour or meal, with respect to its performance in the bakery and the characteristics of bread made from it.

**baklava** — *Greece* and *Serbia/Croatia*. A crisp confection of baked filo layers enclosing chopped or crushed walnuts; it is customarily baked as a large square about 1 inch thick. A sweet syrup is poured over the baked product, before or after it is cut into square or triangular pieces typically about 1 x 2 inches in size.

**bakoom** — *Egypt* disk-shaped, two-layered flatbread made from white flour, water, and salt. Proofed about 90 min before baking.

**bakt** — *Norway* baked.

**bakverk** — *Sweden* pastry.

**balady bread** — one of several middle eastern pocket breads made of sourdough; usual piece size about 150 g.

**balancing** — (1) In flour milling, adjusting the operations of the various breaks and separations in a roller mill to efficiently produce a desired product. (2) In bakery product formulation, adjusting the proportions of the ingredients so as to yield a workable dough and a product having the desired qualities.

**balloon whip** — an agitator for an electric mixer that consists of many wires leading from a top disc around the bottom and back to the top on the other side. Used for whipping egg whites and the like.

**balsams** — resinous exudates of trees or bushes, used as flavors or aromatizers. Similar to gum resin except they contain cinnamic or benzoic acids or their esters.

**banana** — the fruit of *Musa paradisiaca*; when ripe, used sliced or mashed as an ingredient in fillings, frostings, icings, etc. When green, cooked as a vegetable. The banana meringue (or cream) pie is a mainstay of American dessert cuisine. Artificial banana flavors are also widely used in confectionery and baked products.

**Banbury cakes** — *Old English* thin circle (about 3 to 4 inches diameter) of unleavened pastry dough on which is placed a filling of currants, candied peels, spices, butter, sugar, and dried ladyfinger crumbs. The dough is folded over the filling, coated with egg white and sugar, and baked.

**band** — (1) A paper or plastic strip placed around a loaf of bread or other bakery product as part of the package. (2) A metal hoop that is placed on a baking sheet to be filled with batter, sometimes after being lined with paper. (3) The moving continuous horizontal metal strip that forms the baking surface of a band oven.

**band guides** — in a band oven, metal rollers or projections fitted on the framework that push the band toward a central position, used for the purpose of preventing the moving hearth from getting out of line.

**band oven** — an oven in which the hearth is a continuous belt of steel traveling around large cylinders at each end of a tunnel that constitutes the baking chamber.

**band slicer** — equipment for slicing bread loaves and the like; it has sawlike metal bands supported on two rotating drums.

**banh pho** *Vietnam* rice stick noodles.

**banh trang** *Vietnam* thin, semi-transparent, hard ricepaper crepes. They are moistened before use as wrappings for spring-roll fillings and the like. Ingredients include rice flour, water, and salt.

**banh uot** — *Vietnam* similar to banh trang, but made and used when fresh.

**banketletter** — *Netherlands* pastry with an almond-paste filling.

**banneton** — a woven wicker basket in which a yeast-leavened dough is placed to ferment.

**bannock** — *UK/Scotland* flatbread, usually disk-shaped and unleavened. Cooked on a griddle. Variants have been made of oatmeal, barley flour, or wheat flour.

**baps** — the "breakfast rolls of Scotland." Doubtless many versions of this yeast-leavened bun can be found; there are recipes calling for just flour, water, salt, and yeast while others specify eggs, lard, milk, and other enrichments. "Floury baps" are said to be traditional; the dough pieces are brushed with milk (or water), then dusted with flour (this operation may be repeated) immediately before they are placed in the oven. The dough pieces are docked in the center with the forefinger before baking.

**bara brith** — *Welsh* a yeast-leavened sweet bread of rich formula, containing currants and/or raisins, candied fruit peels, etc.; flavored with various spices, such as cinnamon and nutmeg.

**barbari** — *Iran* a Middle Eastern flatbread, having an elongated oval shape with longitudinal ridges on top; it is washed with oil, etc., and extensively docked before baking. The average loaf is about an inch thick and weighs about 700 g.

**barley** — any cereal grass belonging to the genus *Hordeum*, and the grain obtained from these plants. The principal cultivated varieties at present are *H. vulgare*, the six-rowed barleys, and *H. distichum*, the two-rowed barleys. The grain has been used for many centuries for making malt and as a source of carbohydrates in brewing; it is also widely used as an ingredient in soups and foods of some other types.

**barley sprouts** — the shoots formed by moistened barley grains as a early stage in the development of a new plant; collected in large quantities in the manufacture of malt and sold as a animal feed, also in recent years, advertised as a health food.

**barm** — an English/Australian term for the old-style starter cultures and sourdoughs formerly used to leaven bread; the predominant component is yeast. In olden days, barm was often obtained from breweries as foam collected from beer vats.

**barmbrack** — *UK* also, "barnbrack." Rich currant bun or plum cake — a Halloween specialty.

**barometer** — an instrument for determining atmospheric pressure.

**bar presses** — cookie forming machines of the type that extrudes a continuous strip of dough on to a baking pan or oven band.

**barquette** — small boat-shaped pastry shell with either sweet or savory filling.

**barrier properties** — the characteristics of a container or packaging material that prevent the transfer of vapors (oxygen, water vapor, etc.) or liquids (grease, water, etc.).

**barritos** — *Mexico* bar cookies.

**bars** — cookies having oblong or rectangular shapes. Made commercially by extruding dough through a bar press (rout press) machine.

**base** — (1) Bottom portion of a loaf of bread or other object. (2) A chemical compound that yields hydroxyl ions when dissolved in water; approximately the same as an alkali.

**base cake** — the cookie or cookies on which some filling, icing, etc. is applied. The two wafers forming the top and bottom of a sandwich cookie are typical base cakes. Usually, however, a cookie onto which a minor amount of decorative glaze, powdered sugar, or granulated nuts is applied would not be called a base cake.

**base mix** — a preblend of the minor ingredients in a formula; used to improve scaling efficiency and reduce errors at the mixer. The base mix is added to the flour, water, and other major ingredients when doughs are being prepared.

## GLOSSARY OF CEREAL TECHNOLOGY TERMS 15

**basil** — a spice or herb made from the leaves of a annual plant allied to the mint family. The flavor has been described as pungently aromatic, sweet, and spicy. A common constituent of tomato toppings for pizza.

**basmati** — *India* a type of long-grain rice regarded as being superior in its culinary properties to other types of long-grain rice.

**bassinage** — *France* making a dough softer by adding more water during the kneading step.

**bâtard** — *France* a shorter, thicker loaf than the baguette. Typically, a foot in length and 24 oz. in weight.

**batch** — the amount of dough, etc., produced at one mixing.

**Bath bun** — *UK* a light-textured sweet roll, generally round, and usually containing currants. Often topped with sugar. Its name refers to the city in England where it reputedly originated.

**Bath Oliver** — *UK* an old type of cracker or biscuit, stated by Dr. Oliver to be very beneficial to the health.

**battawi** — *Egypt* disk-shaped flatbread made from high extraction wheat flour, water, and salt. Dough baked immediately after mixing. Said to usually include 2 or 3 percent of fenugreek seed.

**batter** — a homogeneous flowable mass prepared by mixing flour and other dry ingredients with water and other liquid ingredients.

**batter beater** — an agitator for an electric mixer; in shape it is a vertically oriented flat paddle having thick arms leading from each side of the top to a central post. It is a versatile agitator, but is most often used for cake batters.

**batter depositers** — pieces of equipment that measure a predetermined amount of (e.g.) cake batter into baking pans.

**batter sponge** — a very soft pumpable ferment or sponge-phase mixture used in some breadmaking processes.

**bauernbrot** — *Germany* rye bread or wholemeal bread.

**Baumé** — a measure of the soluble solids in a sugar syrup, expressed as "degrees Baumé." Often spelled (incorrectly) Beaumé.

**baunilha** — *Portugal* vanilla.

**Bavarian cream** — originally, a dessert consisting of whipped cream folded into a whipped flavored jelly based on gelatin. Now, it often means aerated (or even non-aerated) starch puddings used as fillings for Bismarcks and the like.

**bavarois** — *France* Bavarian cream.

**bazlama** — *Turkey* flat bread made from flour, water, salt, and sourdough. Dough fermented 2 to 3 hr, cut into pieces of about 0.5 lb weight, flattened, and baked immediately on a hot griddle.

**bean jelly** — *Asia* an elastic, rubbery gel made from ground mung beans, powdered rice, or buckwheat and used as small cubes in sauces, as fillings for steamed dumplings, and in other ways. Bland in flavor.

**bean thread vermicelli** — a paste made from mung bean flour and water is extruded in very thin strands, then dried. In appearance, like threads of clear plastic.
**beard** — the small bristles that grow at one end of the wheat kernel.
**beat** — to incorporate air into a batter by using rapid vigorous strokes of an agitator.
**beaten biscuit** — also, "Maryland biscuit." a kind of small cracker made of a lean dough that has been leavened by repeated folding and pounding. Often formed into small disks and baked almost to dryness.
**bebida** — *Spain* drink.
**beer** — an alcoholic beverage usually prepared by fermenting a mash consisting of some malted barley, other cereal preparations, and hops. Many variations exist.
**beeswing** — the two outer layers of the bran coat of the wheat berry, so-called because of their cross markings and light texture.
**beet sugar** — sugar derived from the sugar beet; when refined, consists of about 99% sucrose.
**beignet** — *France* a fritter made with pâte à chou and deep fried; usually filled with fruit, vegetables, or meat.
**beignets soufflés** *France* deep-fried puffed fritters made from pâte à chou.
**belan** — *India* rolling pin.
**beli hleb** — *Serbia/Croatia* white bread.
**belt feeder** — ingredient measuring device consisting of a moving horizontal belt onto which a layer of ingredient is applied, the metering occurring as a result of the speed of the belt and the rate of ingredient deposition; using coupled with electronic measuring circuits.
**belt weighers** — these devices weigh the amount of an ingredient or product resting on a defined area of a horizontally aligned conveyor belt.
**bench** — the table on which the baker manipulates his dough.
**bench brush** — a brush about twelve inches long, with a horizontal handle and vertically aligned soft bristles, used for cleaning the work bench.
**bench tolerance** — the length of time a dough will retain good processing capability after it has been mixed and before it is baked.
**benne seeds** — sesame seeds.
**benzoic acid** — a food preservative generally effective only in acidic environments, such as fruit jellies; not much used in bakeries.
**benzoyl peroxide** — a bleaching agent that has been used by millers to treat flour intended for yeast-leavened products.
**berliner** — *Germany* jam-filled doughnut.
**berliini pannkoogid** — *Estonia* doughnuts.
**berry** — (1) A kernel of wheat; the term is sometimes applied to other cereal grains. (2) One of several kinds of small roundish juicy fruits without stones, e.g., blackberries and raspberries.

**besan** — *India* gram flour, chickpea flour, lentil flour. Used for making batters, doughs, dumplings, and noodles.
**beta amylase** — enzyme that hydrolyzes soluble starch and dextrin to produce maltose; known as the saccharifying enzyme.
**betabel** — *Spain* beet.
**beurre** — *France* butter
**BHA** — butylated hydroxanisole, a synthetically prepared substance that retards the oxidation of fats, thereby slowing the development of rancidity.
**bhajia** — *India* savory fritters.
**bhat** — *Nepal* rice.
**bhaturas** *India* a fried flatbread, circular, used like pita to contain various fillings.
**BHT** — a chemical antioxidant very similar in its action to BHA.
**bialy** — a yeast-leavened roll flavored with onion bits and/or garlic, often made in the shape of a bagel, at other times formed into a flat round bun with a central circular depression.
**biaxially oriented film** — a plastic film (such as polypropylene) that has had its molecular structure highly oriented in two dimensions; this treatment improves many of the film's physical properties, including its transparency and strength.
**bibingka** — *Philippines* fermented rice dough, also a kind of sweet snack made from it.
**bienenstich** — *Germany* cake containing honey and almonds.
**bier** — *Germany Netherlands* beer.
**bihoon** — *Philippines* rice vermicelli.
**bimbollos** — *Mexico* a trade name for hamburger buns with sesame seeds
**bimbuñuelos** — *Mexico* trade-named version of buñuelos, consisting of a fried, octagonal sweet bread with a topping of powdered sugar.
**bin** — a kind of large container used for holding bulk dry ingredients; bins are made in a wide range of sizes and in many shapes.
**bioavailability** — the extent to which a nutrient present in a food can be taken up by a human (or animal) digestive system processing that food and used in the metabolic processes of the organism. For example, the amount of iron in a food, as determined by chemical analysis, may not be entirely available to human consumers because of its chemical structure or the presence of chelating materials (e.g., phytic acid) in the foods.
**biotin** — a vitamin of the B-group, formerly known as vitamin H.
**birra** — *Italy* beer.
**bischofsbrot** — *Germany* a special cake containing a large proportion of fruits and nuts.
**biscocho** — *Philippines* bread toast that has been dried in the oven for use as a thickening ingredient for gravies and the like.
**biscoito** — *Portugal* cookie.

**biscotel** — *Mexico* cinnamon roll-shaped pastry topped with powdered sugar and pecans.

**biscotte** — *France* a rusk made by toasting slices of bread made from a fairly rich dough, yeast-leavened.

**biscotto** — *Italy* rusk, cookie, cracker.

**biscuit** — in the US, a small round flat-topped, soft-textured bread made of a fairly lean dough leavened with baking powder; a baking powder biscuit. In most other English-speaking nations, "biscuits" refers to cookies and crackers, i.e., small thin dough pieces, either sweet or savory, baked to a low moisture content.

**biskota** — *Greece* cookies.

**biskuitrolle** — *Germany* Swiss roll, jelly roll. Could be filled with butter cream icing.

**biskviit rabarbriga** — *Estonia* like a pineapple upside-down cake, but made with rhubarb.

**biskvit** — *Russia* cakes, cookies.

**bismarcks** — doughnut (or other shape of fried yeast dough) filled with creme, jelly, or the like, and often glazed or iced.

**bitter chocolate** — cacao nibs that have been roasted and finely ground; synonyms are bakers' chocolate and chocolate liquor.

**bitter resins** — in brewing, a constituent of hops that has a major influence on the flavor of beer.

**bittersweet chocolate** — an "eating" or ingredient chocolate containing at least 35% chocolate liquor, the remainder being mostly sugar and cocoa-butter with some vanillin, etc.

**bizcocho** — *Mexico* cookie or cracker. *Spain* sponge cake or similar type of cake or bun.

**bizcotela** — *Spain* glazed or coated cookie.

**black bun** — *Scotland* a cake formed from a butter enriched yeast-leavened dough, fairly sweet and containing dried fruits, nuts, etc., that has been wrapped in a moderately thick (perhaps 0.2 to 0.5 inch) layer of a dough not containing fruits and nuts. The loaf may weigh as much as 5 or 6 lbs. Placed in a hoop for baking.

**blackstrap** — a kind of molasses having a very dark color, a strong bitter flavor, and a relatively low sugar content. Used mainly for animal feeds.

**blanch** — (v) to briefly immerse in boiling water or in some other way partially cook the surface of a piece of food.

**blanched nuts** — shelled nuts from which the outer colored skin has been removed by any method.

**blancmange** — a dessert made of milk, gelatine, sugar, starch, flavor, and color. Sometimes with fruit pieces suspended in it. Typically molded in a fancy shape.

**blancs d'oeuf** — *France* egg whites.

**bland** — having little or no flavor; insipid.

**blanquillo** — *Spain* egg whites.

**blas meala** — *Ireland* an alcoholic mixed drink containing some oatmeal (often sprinkled on top of the portion).

**blast freezer** — a room or other inclosed space through which a high velocity current of cold (about -30°F) air is forced so as to quickly freeze the food products contained therein.

**blé** — *France* wheat.

**blé sarrasin** — *France* buckwheat.

**bleached flour** — wheat flour that has been treated with chemicals such as benzoyl peroxide in order to increase its whiteness; usually, some maturing action also results from these treatments.

**bleaching** — (1) Treating flour with oxidizing agents to decolorize some of the natural pigments and/or cause desirable changes in the gluten proteins. (2) In oil processing, a treatment to remove natural pigments and other impurities; the colored substances are commonly absorbed on activated charcoal or, more commonly, bleaching earth or clay.

**bleeding** — losing gas from cut edges of a dough piece.

**blending** — (1) The process of combining lots of wheat or other grain from different bins into a uniform batch meeting predetermined specifications. (2) Making a wet or dry mixture of ingredients, not necessarily either a dough or batter.

**blinchiki** — *Russia* thin, unleavened pancakes.

**blind** — (1) A condition in which the holes in a sieve or screen have been clogged with the material being sifted so that particles can no longer pass through. (2) A blurred design on a cookie, or the like, due to unintended sealing together of patterns that were cut or impressed on the raw dough piece, s result of flow occurring in the oven. (3) Describes a pastry shell that has been baked without a filling.

**blini** — also, "bliny." *Russia Finland* a version of pancakes, usually made with yeast and buckwheat flour, and eaten with sour cream, caviar, etc.

**blintz** — also, "blintze." A thin, crisp, sweet pancake folded over cream cheese, jam, caviar, or other filling.

**blister** — a bump with a cavity beneath it that has formed on the surface of a dough piece during baking or frying.

**blitz method** — a method for making puff pastry that eliminates the step of layering shortening on dough sheets; chunks of puff pastry fat are added to a partially developed dough and then mixed briefly to distribute the chunks fairly uniformly throughout the dough. This is usually followed by sheeting out, folding, and re-sheeting the dough, sometimes repeatedly.

**bloom** — (1) The desirable visual texture and bright color found on the crust of well-baked loaves and rolls. (2) A defect in the appearance of chocolate consisting of a dull gray or white surface coating; it results either from liquid water contacting the surface or exposure of the piece to temperatures near or above its melting point.

**bloom strength** — a measure of the gel-forming strength of a sample of gelatin.

**bocadillo** — *Venezuela* a rich fruit cake made from crystallized fruits and rinds, sold in packets wrapped in banana leaves.

**BOD** — biological oxygen demand; the total amount of oxygen taken up by the microorganisms in a sample of water maintained at specified conditions for a certain extended period of time.

**body** — (1) Response of the crumb to pressure; one aspect of bread texture. (2) Consistency or viscosity of a plastic or semifluid mixture, such as a starch pudding.

**boil** — the processing method whereby enough heat is added to a liquid material so that some of the liquid is at all times turning into vapor and escaping from the surface as bubbles.

**boiled icing** — an icing made by boiling sugar and water to thread stage (238°F) then slowly adding the hot syrup to whipped egg whites with additional beating.

**boiled meringue** — a meringue made by pouring sugar syrup boiled to the hard ball stage (about 250°F) over beaten egg white. Also called Italian meringue.

**boiler** — a vessel designed to furnish a supply of pressurized steam by efficiently and continuously transferring heat to liquid water.

**bola de Berlim** — *Portugal* doughnut.

**bolacha** — *Portugal* cookie.

**bolacha de água e sal** — *Portugal* cracker.

**bolillo** — *Mexico* small (about 2 to 4 oz) football-shaped, yeast-leavened hard or soft bread roll.

**bolle** — *Denmark* bun.

**bollito** — *Spain* roll or bun.

**bollo** — *Spain* biscuit or small cake.

**bolo** — *Portugal* cake

**Bologna-style pasta** — alimentary pastes formed into fancy shapes, such as "bow-ties."

**bolting** — sifting, esp., particle classification in a flour mill.

**bombe** — a molded (usually dome-shaped) dessert constructed from several different types of ice cream; it is cut in portions immediately before serving, and often topped with a sauce. The term has been applied, infrequently, to dome-shaped cakes.

**bonbon** — a small candy piece, usually a soft center coated with chocolate or similar enrobing material.

**börek** — *Turkey* savory pastry made out of thin layers of dough and butter, with various kinds of cheese or meat filling.

**Boston brown bread** — a dark sweet bread (chemically-leavened) containing cornmeal and molasses, and often raisins and spices, among other ingredients; often cylindrical. Traditionally steamed, not baked.

**Boston cream pie** — two cake layers (usually white) separated by a thick layer of vanilla pastry cream (or starch pudding) and frosted with chocolate icing.

**boter** — *Netherlands* butter.

**bottom heat** — in an oven, the heat that is transferred to the product (or the pan) from the hearth, by conduction.

**botulism** — acute food poisoning caused by a toxic chemical originating from *Clostridium botulinum*; often fatal.

**bouchée** — *France* a shell of puff pastry sized for one serving, such as a patty shell.

**bouchée à la reine** — *France* puff-pastry shell filled with meat, sweetbreads, seafood, or mushrooms.

**boulangère** — *France* baker, esp., a baker of bread-type products.

**boulangerie** — *France* bakery.

**bound water** — water molecules that have their water-like properties chemically or physically reduced by other substances (such as hydrophilic colloids) in their environment.

**bowl extension** — a metal or plastic ring or hood placed around the top of a mixer bowl so as to reduce the amount of material thrown out of the bowl during violent agitation.

**bowl knife** — a plastic spatula or flexible dull-edged knife used to scrape batter or dough from the sides and bottom of a mixer bowl.

**boxty bread** — *Ireland* a broad category of thick cakes consisting (always) a high proportion of potatoes mixed with enough flour to make a cohesive dough, then cooked on a griddle.

**boza** — *Turkey* thick white liquid ("malted millet") made from milk, millet, breadcrumbs, sesame, and cinnamon that has been fermented and cooked.

**brack** — *Ireland* a yeast-leavened (barm brack) or soda-leavened (tea brack) bread, lightly sweetened and often containing currants, citrus zest, and spices, and cooked in an oven.

**brake** — to pass through a dough brake; a dough brake (q.v.).

**bran** — in grain milling, the fraction or mill streams consisting mostly of the fibrous outer layers of the kernels, formerly used almost entirely in animal foods, now a constituent of many high-fiber food formulas.

**brandy** — an alcoholic liquor distilled from wine, sometimes used as a flavoring agent in bakery foods or their adjuncts.

**brandy snap** — *UK* gingersnap flavored with brandy.

**bran flakes** — a prepared breakfast cereal made from a dough consisting mostly of cereal bran with some sort of binder that is formed into pellets, then into flakes, and toasted.

**bran muffins** — a chemically leavened quick bread in cupcake shape and containing some bran; usually colored dark brown with caramel coloring and often contain raisins.

**brasno** — *Serbia/Croatia* flour.

**braunschweiger kuchen** — *Germany* rich cake containing fruit and almonds.

**Brazil nuts** — a large, rather bland-flavored nut sometimes used in sliced or chopped form as a topping for sweet bakery goods.

**bread** — in a narrow sense, baked foods made of a developed dough containing flour, water, yeast, and salt, and usually containing malt, shortening, and milk. May contain other ingredients as well. Typically, the dough is fermented and otherwise processed before baking to give a low density loaf or roll with a clearly defined, relatively dark crust and a soft, silky light-colored crumb. In a general sense, almost any food containing a considerable amount of grain flour that has made into a dough by mixing with a liquid, then then baked. See also, specific kinds of bread, such as crispbread, quickbread, and flatbread.

**bread bagging equipment** — machines for inserting loaves into pre-formed plastic bags that are generally closed by twist ties or plastic tags.

**bread crumbs** — (1) Small particles generated when a bread loaf is sliced. (2) Ground dried bread used as a coating for fried foods; a similar product is made by an extrusion process not involving fermentation or molding.

**bread faults** — ways in which a specific loaf of bread deviates from predetermined specifications and standards.

**bread flour** — generally, a hard wheat flour of at least moderately high protein content that has been milled and blended so as to make it particularly suitable for processing into bread.

**bread improver** — any of the various compounds and mixtures thereof that compensate for some of the inadequacies of a less-than-optimum flour intended to be used in breadmaking.

**breading** — a coating for foods that are to be fried (sometimes baked), consisting usually of a first layer of an adhesive material based on milk or eggs, then a layer of bread crumbs, cornmeal, or the like.

**bread scoring** — a system of evaluating bread that consists of applying numerical scores to various quality features of the loaf, then summing these individual scores to give an single-figure characterization of the overall quality of the product.

**breadsticks** — relatively thin and long cylinders of bread dough, often (but not necesarily) baked to a crisp stage and sometimes flavored with garlic, butter, spices, etc. Seldom more than one ounce in weight.

**break** — (1) One of the first steps by which the grain is reduced to meal in roller milling processes; usually performed by pairs of grooved steel cylinders. (2) See "break and shred."

**break-and-shred** — The portion of inner crust exposed when the outer crust ruptures during oven spring; in a pan loaf of bread, it is the lighter and rougher area along the side of the loaf just above the pan top.

**breakdown** — (1) The development of undesirable chemical or physical changes in a frying fat. May include darkening, formation of excess free fatty acids or peroxides, polymerization and gumming, undesirable foaming, and development of unpleasant odors and flavors. (2) The complete mechanical failure of a piece of equipment or an entire production line.

**breakfast cereals** — an extremely varied category of foods that includes both grains processed only slightly and requiring cooking before consumption and grains that have been highly modified by flaking, extruding, puffing, etc. These foods are most often used as the first meal of the day in combination with milk and sugar, but are also consumed as snack products and utilized as ingredients in other products.

**break flour** — flour produced by the break rolls as an incidental component when the grain is torn apart in the break operation.

**breaking down** — deteriorating of the physical properties of a dough, creamed mass, or batter due to excessive mixing.

**break system** — in a flour mill of the conventional modern type, a series of pairs of corrugated steel rollers that tear open the wheat kernels and scrape endosperm from the bran.

**brei** — *Germany* porridge, mash, puree.

**brezel** — *Germany* salted pretzel in the traditional shape.

**brew** — a bread dough intermediate consisting of water, yeast, yeast nutrients, a buffering agent, and often some flour, that has been fermented for a specific time at a predetermined temperature. After fermentation is complete, the brew is mixed with the remainder of the ingredients, including the flour. A brew serves many of the same purposes as a liquid sponge or a "ferment," but generally contains less flour and more water and is less viscous.

**brewers' dried grains** — the insoluble material remaining in the brewer's cooking vat after draining off the wort; it is dried and used as animal feed.

**brewers' grits** — in the dry milling of corn, a particle size range of the corn endosperm that is especially well-suited for use in the brewing process.

**brewing** — the entire process of making beer.

**brewing by-products** — the leftovers from the beer-making process, consisting mostly of insoluble grain materials collected from the cooking vessels, typically dried and used as ingredients in animal feeds.

**brine** — usually, water that is saturated (or nearly so) with sodium chloride. In pickling of meats, brine may include other important ingredients such as nitrates or nitrites.

**brioche** — this term has been applied to a rather wide range of baked rolls and loaves; brioches are typically made from yeast dough and are rich in butter and eggs but are not particularly sweet. Often formed into large rolls or small loaves consisting of a larger bottom portion with a smaller "topknot." Sometimes, they are made with a center filling of fruit.

**Brix** — a scale for converting the specific gravity of a syrup into its sugar concentration. Gives only an approximation of the true sugar content if other solubles are present.

**broa** — *Portugal* (1) A thick cracker based on cornmeal. (2) A kind of gingerbread.

**brochan** — *Scotland* porridge and especially oatmeal porridge.

**bröd** — *Sweden* bread.

**brod** — *Denmark* bread.

**bromated flour** — flour to which the oxidizer or maturing agent potassium bromate has been added.

**bromates** — the oxidizing or maturing agents potassium bromate and sodium bromate.

**bromelain** — a protein-digesting enzyme obtained from pineapples and sometimes used to reduce the mixing time of doughs. Has been largely or entirely replaced by enzymes obtained from bacterial or fungal cultures.

**brood** — *Netherlands* bread.

**brood pudding** — *Netherlands* a pudding made from bread mixed with eggs, cinnamon, and rum flavor.

**broodje** — *Netherlands* bread roll.

**broomcorn** — a type of *Sorghum bicolor* (also, *Sorghum vulgare*), the panicle of which has been used for centuries to make brooms for sweeping.

**brose** — a Scottish preparation made by stirring hot water, milk, or broth into, e.g., oatmeal until it forms a thick porridge. Oats in some form is usually the principal constituent, but barley or other grain meals have also been used. Athol brose is a drink of honey, oatmeal, water, and whisky.

**brot** — *Germany* bread.

**brotchan foltchep** — also, "brotchan roy." *Ireland* A soup made principally of oatmeal, leeks, and milk, often slightly spiced.

**broth** — in bread-making, a fermenting liquid prepared from yeast, water, yeast nutrients, and sometimes flour. It is used to replace bulk fermentation or to greatly reduce the duration of the bulk-fermentation step.

**brown-and-serve** — describes bread loaves or rolls that have been baked until the crumb "sets up" but not long enough to brown the crust; the consumer bakes the product in a home oven until optimum crust color is obtained.

**brown betty** — a pudding of apples and bread crumbs deposited in alternating layers and baked.

**brown bread** — has been applied to many kinds of yeast-leavened bread and a few types of chemically leavened breads, most of them being brown in color, slightly sweet, and containing whole wheat.

**brownie** — a moist, chewy, dense, chocolate cookie baked in sheets and cut into rectangular pieces. Occasionally used to refer to cookies of similar physical characteristics but not containing chocolate, e.g., butterscotch brownies. Frequently iced or frosted, sometimes contain nuts.

**browning reaction** — the so-called Maillard reaction, which involves the interaction of amino acids and proteins with reducing sugars; it produces brown-colored poorly defined compounds that often have pronounced flavors. It is the principal cause of crust coloration.

**brown sugar** — granulated refined sugar, the particles of which have been coated with cane molasses. The refined sugar portion can be either cane or beet sugar. Contrary to general belief, the brown sugar of commerce is not a partially refined sugar.

**brun kage** — *Denmark* brown, spicy cookie.

**brunt farinsocker** — *Sweden* brown sugar.

**Brussels biscuit** — a form of zwieback.

**brynt** — *Sweden* browned.

**bstila** *Morocco* a pie made from flaky puff pastry and having a filling of scrambled eggs and chopped pigeon meat.

**BTU** — or, "Btu." British thermal unit, the amount of heat required to raise the temperature of one pound of water one degree F. Equals 0.252 Cal.

**buckle** — (n) a sweetened fruit mixture covered with a biscuit dough on the top of which is deposited a streusel-type mixture; the whole structure is baked until the dough is completely cooked and the streusel browned.

**buckweisen** — *Germany* buckwheat.

**buckwheat** — the fruit (grain), of an annual herb native to Siberia. Dark in color, roughly triangular in outline. A grain but not a cereal. Common buckwheat is *Fagopyrum esculentum*.

**bucky** — a bucky dough is tough and dense and resists extension; it tends to tear when stretched. Buckiness is characteristic of a "young" or under-fermented dough, but there are other causes, as well.

**budding** — one of the reproductive processes in yeast and some other microorganisms; it is characterized by the formation of a protuberance on the outer wall of a cell; this later expands until it is finally cut off from the original cell and begins a separate life.

**budin** — *Spain* a sort of cake or trifle, served like a pudding.

**budino** — *Italy* a kind of dumpling for use in soups and the like.

**buffering agent** — also, "buffer." A material that, when present in an aqueous system, tends to reduce the pH change caused by addition of acids or alkalies to the system.

**bulgur** — also "burghul," "bulghur," "bulgar," etc. Grains of wheat that have been soaked and heated until the starch gelatinizes, then dried and the bran layers removed. Typically cracked into relatively small granules and used in cooking somewhat like rice.

**bulk handling** — the practice of receiving, transferring, storing, and dispensing of ingredients not contained in bags, cases, drums, etc.

**bulk fermentation** — the stage in the baking process in which the dough is fermented in the condition it is removed from the mixer, i.e., before it is separated into pieces.

**bulking agents** — substantially inert ingredients that are added to a mixture primarily to increase its weight or volume.

**bun** — see "buns." *Vietnam* rice stick noodles.

**bun divider** — a device used mostly in retail shops for dividing a dough mass into pieces of uniform weight by first pressing the mass into a sheet of uniform thickness and then severing pieces by a number of automatically activated knives. Often combined with a rounding device.

**bun machines** — generally, a production line assemblage consisting of a dough hopper, divider, and rounder, the latter often of a belt-and-channel design.

**bun pan** — a sheet pan provided with many cups or shallow cavities to restrict shifting of dough pieces during movement of the pan.

**buns** — small (8 oz. or less) pieces of baked bread dough, sometimes in fancy shapes and/or with enriching ingredients. *UK* a type of sweet roll, flavored with fruit, preserves, coconut, etc., but not in fancy shapes.

**buñuelos** — *Spain* fried cookies in thin shapes; in one form, made from a fluid batter coated onto fancifully shaped metal molds and then deep-fried; they are often finished by coating with sugar and cinnamon. Also, the term has been applied to a fried wheat flour tortilla coated with sugar and cinnamon.

**burek** — *Serbia/Croatia* a meat or cheese pasty.

**burma** — *ME* see "kadaifi."

**burning in** — also, "burning out," A process applied to new baking pans for the purpose of conditioning their surface so they will absorb heat better and release the baked product easier. Typically requires greasing the pan and then heating it at 400°F for 30 min or more.

**burr** — *ME* Arabic bread made from whole wheat meal.

**burrito** — *Mexico* a soft tortilla (either wheat or masa) folded around and completely enveloping a filling of mashed beans, spiced meat, etc.

**burro** — *Italy* butter. *Mexico* a kind of banana.

**bursting** — wild or uncontrolled break-and-shred resulting from excessive oven spring.

**burtonizing** — the addition of hardening minerals (principally calcium salts) to brewing water to reduce the extraction of unwanted materials from the malt during mashing.

**bushel** — a measure of dry volume used for grain, etc., traditionally described as 4 pecks or 32 quarts; it has been standardized (Winchester bushel) at 2,154.02 cubic inches, or (US standard bushel) 35.2383 liters.

**butter** — fatty ingredient obtained by churning sweet or sour cream from cow's milk; used as a shortening, wash, etc., in premium bakery foods.

**buttercream frosting** — rich uncooked frosting containing at least powdered sugar and butter (or other shortening) whipped to a smooth, uniformly aerated condition.

**buttercream icing** — not clearly differentiated from buttercream frosting, but probably generally understood to be less aerated than frosting; there are also some cooked forms.

**butterdejg** — *Denmark* puff paste.

**butterfat** — the lipids that constitute approximately 80% of butter; same thing as milk fat.

**butterflake rolls** — individual bread rolls typically made by spreading a thin layer of butter on a sheet of yeast-leavened dough, cutting the dough into strips about 1½ inch wide, folding a length of the strip into about 5 layers, inserting these into a cup of a muffin tin (with strips vertically oriented, cut ends on top), then baking. Other methods are being used, including completely automatic systems not requiring a prior sheeting step.

**butter horns** — basic sweet dough that has been cut and shaped like horns; contains some butter or is brushed with butter.

**butter icing** — a creamed mixture of butter and powdered sugar with other ingredients, forming a rich smooth oleaginous frosting or icing. Commercially, some of the butter is usually replaced with margarine or an appropriate shortening.

**buttermilk** — the fluid remaining after sour cream has been churned and the butter removed; this is the material that forms the basis for the commercially available dried buttermilk sometimes used as an ingredient in bakery products. Cultured buttermilk, as found in the supermarket dairy case, is generally skim milk in which bacterial cultures have been allowed to grow until the desired physical and chemical changes have occurred.

**buttermilk bread** — yeast-leavened white bread containing a specified amount of buttermilk solids, usually in the form of spray-dried buttermilk.

**butters** — also, "fruit butters." A term sometimes applied to fruit jams having the fruit present in the form of fine particles; apple butter is by far the most common variety and it is often highly spiced with cinnamon. The principal (if not only) use is as a table spread for bread and toast.

**butterscotch** — a candy made of brown sugar and butter cooked to a very low moisture content until the typical appealing flavor forms; also natural and artificial flavors simulating the flavors obtained in such a process.

**butter sponge cake** — cake made from sponge cake batter to which butter or other shortening has been added.

**by-product** — a material that inescapably results from the production of a desired food, e.g., whey is a by-product of cheese manufacturing.

**byezye** — *Russia* meringue.

## -C-

**cabinet proofers** — chambers or other forms of enclosures in which the dough product in its final shape (often in a baking pan) is retained for a period of time sufficient for the final proofing to occur.

**cabinet pudding** — *UK* a pudding made of bread or cake combined with eggs, candied fruit, and milk; usually served hot and topped with a sauce.

**cacahuates** — *Mexico* peanuts.

**cacao products** — chocolate, cocoa, cocoabutter, and other products made from the cocoa bean.

**cadaif** — see "kadaifi."

**caju** — *Portugal* cashew nut.

**cake** — a soft sweet baked product of widely variable size made from a flour-containing batter and often frosted and filled. Can be leavened with chemicals, steam, or air, or combinations of these.

**cake beater** — the agitator blade used primarily for mixing batters in vertical mixers. It consists of a moderately thick blade shaped to conform closely to the bowl sides and having large openings.

**cake des rois** — *France* there are various types of this yeast-leavened "king's cake." Most of them are made in multi-portion ring shapes. The Bordeaux variety is, or was, made from a dough containing flour, yeast, butter, sugar, salt, eggs, and water, sprinkled with sugar crystals, garnished with candied citron slices, and brushed with orange-flower flavored syrup.

**cake flour** — a wheat flour of low protein content, having a very white color and fine particle size; often treated with chlorine.

**cake hoop** — a circular metal ring, one to a few inches in height, set upon a baking sheet to contain cake batter during baking.

**cake liners** — paper or parchment cups into which cake batters are deposited prior to baking; also the paper discs or sheets used in the bottoms of baking pans, whether circular or sheet.

**cake pans** — containers of metal or oven-resistant plastic into which batters are scaled and then carried through a baking chamber.

**cake rolls** — thin layers of any kind of cake, spread with a filling (jam, marshmallow, whipped cream substitute, etc.), then rolled up so that a cross-section reveals a spiral pattern of alternating cake and filling; jelly rolls and swiss rolls are examples.

**cake slabber** — a cutter for horizontally slicing sheet cakes or the like into thinner layers.

**calabaza** — *Spain* pumpkin, sometimes squash.

**calcium acetate-propionate** — a substance containing a mixture of compounds, among them calcium propionate and calcium acetate, that has been recommended for adding to dough as an inhibitor of mold growth in the baked product.

**calcium carbonate** — a mineral substance used as a calcium source in yeast foods and as an enriching agent in bread, etc. An effective buffer.

**calcium peroxide** — a dough oxidant having somewhat different properties than the bromates or iodates.

**calcium phosphate** — there are a number of different calcium phosphates, some useful as leavening acids, nutritional supplements, buffers, etc.

**calcium propionate** — a mold inhibitor.

**calcium stearoyl lactylate** — a dough conditioner that will, in many cases, improve the processing response of doughs and batters and extend shelf life of the finished product.

**caldo** — *Italy* hot.

**caliente** — *Spain* warm, hot.

**calzone** — *Italy* a folded pizza shell with all the "topping" inside; it can be baked or deep fried.

**camotes** — *Mexico* sweet potatoes. sometimes used like pumpkin to make sweet spiced fillings for pastries.

**campechanas** — *Mexico* a pastry made of a strudel-like dough sheet which is rolled into a tube containing a flavored paste, cut crosswise, sprinkled with sugar, and baked.

**canache** — *France* a chocolate filling made of cream and chocolate liquor or other couverture, plus other ingredients. Also known as "Paris cream."

**Canadian cheese** — when it refers to a bakery product, this means a pastry made of cooked meringue filled with a marron-flavored butter cream.

**canapé** — *France* a small piece of bread or toast topped or filled with a savory meat paste, caviar, cheese, or the like, served as an appetizer.

**Candida crusei** — a yeast found in some sourdough starters that have been used in Germany.

**candied ginger** — a piece of the root or rhizome of the ginger plant that has been treated with hot concentrated sugar syrups until the tissues have been thoroughly impregnated with the sweetener.

**candied peels** — orange or lemon peel (rarely lime, sometimes citron) that has been impregnated with saturated sugar syrup.

**candying** — modifying a fruit piece by replacing its internal fluids with a concentrated sugar syrup. This treatment greatly changes the texture, flavor, and appearance of the fruit, but makes it resistant to microbial spoilage. In the case of candied cherries, only the shape and basic structure of the fruit remain, the color and flavor having been replaced with additives.

**canederli** — *Italy* dumplings made from breadcrumbs, ham, and sausage.

**canela** — *Portugal Spain* cinnamon.

**cannella** — *Italy* cinnamon.

**cannelle** — *France* cinnamon.

**cannelloni** — *Italy* another form of pasta, being large tube shapes that can be stuffed with meat, etc., then covered with a sauce and baked.

**cannolo** — *Italy* rolled pastry filled with e.g., cheese, nougat, and crystallized fruits,

**canning** — the process of packaging foods in hermetically sealed metal cans; it usually involves the application of high temperatures to the sealed can for the purpose of reducing or eliminating microbial contamination, thereby increasing the shelf-life to months or years. Some bakery foods have been satisfactorily packed in this manner.

**canola oil** — this item of commerce used to be called rapeseed oil, but a new name was chosen for public relations purposes. It is high in monounsaturated fatty acids. Ordinary rapeseed oil is high in erucic acid, which has some undesirable health implications, but the canola designation is said to be restricted to oil from rapeseed cultivars that are low in this component.

**canthaxanthin** — one of the carotenoid compounds class of synthetic chemical compounds that can be used to color foods.

**CAP** — see "controlled atmosphere packaging."

**capillary melting point** — the temperature at which a fat becomes completely clear and liquid inside a capillary tube of specific dimensions.

**cappelletti** — *Italy* small ravioli with savory fillings.

**capsaicine** — an alkaloid found in cayenne pepper, and probably in other kinds of pepper, that is chiefly responsible for the hot taste of the spice.

**captive bakery** — a bakery operated by a supermarket chain for the purpose of supplying baked goods to the retail outlets.

**caramel** — (1) Caramel color. (2) A firm but usually plastic confection formed by heating a mixture of sugar or other sweetening materials, milk products, fatty materials (such as butter), colors, flavors, and (sometimes) pieces of nuts or candied fruits. Often packaged as small cubes.

**caramel buns** — sweet dough pieces (usually of the cinnamon-roll type) baked in a pan containing cups that have been coated inside with a sugar and shortening mixture.

**caramel color** — a dark brown, nearly flavorless syrup made by processing corn syrup; sold as an ingredient for coloring rye breads, etc.

**caramel corn** — popped corn having the kernels glazed with a thin coating of a solidified, very low moisture syrup based on sugar or corn syrup.

**caramel flavor** — a flavoring material originally produced in much the same way as caramel candy.

**caramelized sugar** — dry sugar heated with constant stirring until it becomes dark in color; used for flavoring and coloring.

**caramellato** — *Italy* caramelized.

**caramelo** — *Spain* caramel. In the plural, small candy pieces, not necessarily caramels.

**caramel rolls** — sweet rolls of the cinnamon roll shape baked in a pan that has had its interior coated with a paste made from sugar, butter, nuts, and other ingredients.

**caraway** — small crescent-shaped grayish-tan seeds harvested annual of the parsley family (*Carum carvi* L.). They have a characteristic agreeable odor, and an aromatic, pleasant, warm, sharp taste. Whole and ground seeds are very frequently used in rye breads and less frequently in cookies.

**carbohydrase** — enzyme that splits carbohydrate polymers into smaller units.

**carbohydrate** — a chemical compound composed of carbon, hydrogen, and oxygen, generally in the approximate atomic proportions of one C, two H, and one O. Examples are starch, sugar, and cellulose.

**carbon dioxide** — a colorless, tasteless, edible gas that is the principle leavening agent of most bakery foods. In these products, it is derived from yeast fermentation or the addition of sodium bicarbonate or ammonium bicarbonate, with a few exceptions.

**carbon filters** — tubes or beds consisting principally of granules of activated carbon, used primarily to remove colored and flavored materials from potable water.

**carboxymethylcellulose** — a modified form of cellulose that binds large amounts of water and is sometimes used as a thickener in batters, icings, frostings, etc.

**cardamon** — decorticated cardamom seed is the dried, ripe fruit of *Elletaria cardamomum*. The hard, wrinkled, light reddish-brown seed has a pleasant aromatic odor and a characteristic warm, slightly pungent taste. This spice, which is relatively expensive, is occasionally used in Danish pastries, cookies, pumpkin pies, and coffee cakes.

**carmine** — a red coloring material extracted from the bodies of certain insects, also a hue resembling that of the pigment.

**carob bean** — also, "locust bean," "St. John's bread," etc. The sweet seedpod of the carob tree. Contains a high content of sugar. Has been roasted and otherwise treated to yield a substitute for cocoa, though it has none of the appealing chocolate aroma and contains very little fat.

**carotene** — a yellow pigment occurring in many vegetables and cereals.

**carotenoids** — natural and synthetic orange and yellow pigments of the same general chemical type as carotene.

**carrageen** — also, carrageenan. A water-binding and gelling agent extracted from certain kinds of seaweeds. It consists of a mixture of compounds, each of which has somewhat different gelling and thickening properties. Used as an ingredient in jelly-type fillings, in frostings, etc.

**carrot cake** — a layer cake formulation of the spice cake type, containing a significant proportion (up to about 10%) of carrots (usually raw, sometimes blanched) in small particle size (shredded, grated, etc.).

**caryopsis** — a small, dry, indehiscent fruit, with a thin membranous pericarp adhering closely to the seed, so that the fruit and seed are combined in one body, forming a single grain, as in wheat, corn, and barley.

**casein** — the predominant type of protein in milk, and which forms the basic structure of almost all kinds of cheese.

**cashews** — nuts harvested from *Anacardium occidentale*, a tree grown in the West Indies, South America, and the Far East. Most cashews entering the international market have been processed in India. They are crescent-shaped seeds, high in fat, mildly flavored, only moderately crisp when roasted, and light in color. Used occasionally as a topping on baked goods, but much less important in U.S.A. commerce than almonds, pecans, walnuts, and peanuts.

**cassata** — *Sicily* sponge cake layers sprinkled with rose water, then spread with a filling based on ricotta sheese, sugar, sour cream, chocolate, nuts, and preserved fruits. The assembled layers are frosted with a cooked white icing. The word "cassata" is also used to describe a type of ice cream.

**cassia** — specifically, the dried bark (usually sold in ground form) of the evergreen tree *Cinnamomum cassia*; now used to mean practically the same thing as "cinnamon," though the latter term was traditionally restricted to material from *Cinnamomum zeylanicum* (Ceylon cinnamon).

**castagnaccio** — *Italy* chestnut cake with pine kernels, raisins, and nuts; cooked in oil.

**castaña** — *Spain* chestnut.

**castanha** — *Portugal* chestnut.

**castera** — *Japan* sponge cake.

**caster sugar** — *UK* also, castor sugar. An imprecise term fairly common in the UK, but seldom if ever used in the US. Essentially the same as granulated cane or beet sugar, though the particle size distribution may be somewhat different (finer in caster sugar). The name originated as a description of the kind of sugar that was placed in the shaker (caster) that stood on the dining table or kitchen table.

**casting sugar** — high boiled (low moisture) sugar syrups used for making molded decorative figures.

**castle pudding** — *UK* an individual sponge cake topped with (or served with) a jam sauce.

**castor sugar** — see "caster sugar."

**catalase** — a widely distributed enzyme that greatly speeds up the decomposition of hydrogen peroxide and, perhaps, of other compounds.

**catalyst** — a material that speeds up a chemical reaction without undergoing permanent change itself.

**cebollas** — *Spain* onions.

**celery seed** — a spice made from the dried fruit of an herb that is related to, but not identical with, the vegetable (stalk) celery plant. The tiny brown seeds have a flavor similar to celery stalks. They have been used in croutons and stuffing mixes, savory chip coatings, pizzas, and snack crackers.

**celiac disease** — also, "coeliac disease." A chronic intestinal disorder mostly affecting young children and caused, at least in some cases, by wheat protein, specifically by gluten. The disease is sometimes fatal, but can often be completely cured by eliminating cereal proteins from the diet.

**cellophane** — flexible transparent film made of regenerated cellulose; has excellent optical properties, but has poor resistance to moisture transfer and poor sealability. Often combined (laminated) with other films.

**cellophane noodles** — a western term for the semi-transparent dried strands (size and shape similar to vermicelli) used in several Asian countries in the same way as spaghetti, but made of bean paste.

**cellulase** — an enzyme that can break down cellulose into smaller molecules such as glucose.

**cellulose** — the chief structural material of all woody plants; chemically, it is a polymer of glucose, like starch but with the glucose residues joined with a different bond orientation. Much more resistant to chemical attack than starch, and not digestible by humans. A dietary fiber.

**cellulose acetate** — a transparent thermoplastic material made by esterifying cellulose with acetic anhydride and acetic acid. Packaging films are made from it, and it can be molded, extruded, and cast into various shapes.

**C-enamel** — a coating applied to the inside of metal cans to reduce the corrosive action of sulfur compounds (from the food) on the metal.

**centeno** — *Spain* rye.

**centrifugal** — in a flour mill, a rotating bolter in which the ground material is forced against a circumferential sieve by rotating beaters while the sieve also rotates.

**centrifuge** — a device that spins liquids at high speeds so as to cause separation of the liquid's components according to their density.

**cereals** — the seeds of grassy plants as used for food or feed; also, the plants yielding such seeds.

**Cerelose** — a trade name for crystalline dextrose.

**cereja** — *Portugal* cherry.

**cereza** — *Spain* cherry.

**certified color additives** — certain chemically synthesized dyes and their lakes that have been approved by the FDA for use in foods if the manufacturing batches have been certified. There are many "uncertified" coloring agents that are also FDA-approved.

**cerveja** — *Portugal* beer.

**cerveza** — *Spain* beer.

**chabacano** — *Spain* apricot.

**chaff** — the fine, light material that separates from grain during harvesting or cleaning.

**chain oven** — a continuous traveling-tray oven in which the trays carrying the product are pulled forward by dogs (projections) on parallel chains. The chains form an endless belt passing through the oven from front to back and then from back to front underneath the baking chamber.

**chalkiness** — a dull, whitish appearance of (especially) rice kernels, sometimes occurring in patches, as contrasted to a clear, vitreous, translucent endosperm.

**champorado** — *Philippines* a chocolate-flavored sticky-rice dessert.

**chamucos** — *Mexico* pastry made by forming a ring of pan fino around a center plug of flavored paste, then baking.

**chapati** — *India* a type of unleavened flat dough piece made from coarse whole wheat meal and baked on a griddle.

**chapelure** — *France* bread crumbs.

**charlotte russe** — a dome shaped mold lined with ladyfingers and filled with bavarian cream.

**chausson** — *France* (1) A puff pastry dough piece garnished with apple jelly before baking. (2) A sweet or savory calzone.

**chausson aux pommes** — *France* apple dumpling.

**chaval** — also, "chawal." *India* rice.

**chateau** — *France* a wine foam served warm as a dessert or as a sauce for puddings, cake slices, etc.

**cheat** — *Old English* plain bread made from a flour containing more bran than white flour does but more refined than whole wheat flour.

**checking** — a serious quality defect found in thin, dry baked products such as soda crackers; it appears as irregular cracks that often extend completely through the thickness of the object.

**cheddar** — a very common type of cheese made from whole milk by the cheddaring process and matured or aged for varying periods of time.

**cheena chatti** — *Sri Lanka* a wok-like pan used for cooking appas and some other types of bread.

**cheese** — a food made by coagulating the casein of milk and then separating the curds (which contain other milk components) from the by-product, whey. The curds are usually pressed into a solid mass, then cured or aged for weeks or months. Among the many variations are inoculation with bacterial or fungal cultures, adding colors, flavors, and salt, and curd washing.

**cheese cake** — there are many different kinds of cheese cake, but all of them contain a substantial proportion of some kind of relatively bland flavored cheese, such as bakers' cheese. Otherwise, there are hardly any unifying factors agreed upon by the experts.

**chef** — a culinary artist; professional cook.

**chelating agent** — a chemical compound that forms unusually stable complexes with metal ions.

**Chelsea bun** — *UK* sweet roll containing dried fruit (such as currants) and glazed with honey.

## GLOSSARY OF CEREAL TECHNOLOGY TERMS

**chemical leavening** — usually some form of baking powder, i.e., a mixture of sodium bicarbonate with an acid reacting substance. Ammonium bicarbonate is also in this category.

**chemically leavened** — a baked product that owes some part of its volume increase to the evolution of carbon dioxide from sodium bicarbonate or ammonium bicarbonate that has been used as an ingredient in a dough or batter.

**cherries** — small stone fruits of certain species of trees or shrubs of the genus *Prunus*. There are sweet cherries, such as the Bing, as well as the red sour pie (RSP) cherries that are commonly used in jams, jellies, preserves, and pie fillings.

**chess pie** — originally a cheese pie, now there are many kinds, most of them not containing any cheese. Probably the most common type has a filling of a kind of custard made with butter, eggs, and sugar, often containing or topped with fruits, pecans, coconut, or the like. The crust is usually made from an unleavened dough of flour, fat, and salt.

**chestnuts** — the sweet edible nut produced by trees of the genus *Castanea*. There are European and American types of this nut. The nut, as used, is relatively high in moisture and low in fat compared to other tree nuts.

**chicha** — *Mexico* originally an Aztec/Mayan/etc. drink made from the fermentation of corn that has been chewed and spit out (adds starch-digesting enzymes). Now, applied to various kinds of alcoholic and non-alcoholic beverages made by, perhaps, more sanitary methods.

**chicharrones** — *Mexico* fried puffed pork rind pieces, used as a snack.

**chichimanga** — also, other spellings. *Mexico* essentially a deep-fat fried burrito, usually fairly small.

**chiffon cakes** — products baked from batters resulting from combining a whipped mass of whole eggs, flour, sugar, and other ingredients with melted butter (or vegetable oil). These cakes have a more open texture and a softer consistency than most pound cakes.

**chilaquiles** — *Mexico* fried tortilla chips stir fried with eggs, meat, and chiles, and served with a topping of chili sauce.

**chili** — hot peppers; any one of many edible varieties of the fruits of *Capsicum annum* or *C. frutescens*. Used as a flavorant in many kinds of foods, including some snack products. *Mexico* The same ingredient, used as a flavor, coating, and color on most anything, including fruits, vegetables, and confections.

**chill** — (n) the hardened external surface of a chilled-iron roller.

**chlorination** — (1) Adding small amounts of chlorine gas to wheat flour in order to whiten it and improve its quality for cake baking. (2) Treating, e.g., water, by adding carefully metered amounts of chlorine gas for the purpose of killing microorganisms and inactivating viruses. A common water treatment process.

**chlorine** — an element that exists in gaseous form when at room temperature and atmospheric pressure. The gas is greenish-yellow, of relatively high density, and with a pungent, acrid, disagreeable odor. It is poisonous even at low concentrations.

**chlorine dioxide** — a gas used to treat (oxidize) flour.

**choclo** — *South America* corn cooked on the cob.

**chocolate chips** — small pieces of some form of chocolate (usually sweet or milk chocolate), often made commercially by depositing small amounts of melted chocolate on a cooled steel band, then tempering the bits.

**chocolate cookie crusts** — a pie or tart crust formed by pressing in a mold a mixture of shortening and coarsely ground cocoa-flavored cookies of the snap type, sometimes with other ingredients.

**chocolate enrobers** — machines that apply a coating of liquid chocolate to a food piece such as a cookie; enrobers are usually combined with tempering equipment to properly condition the liquid chocolate.

**chocolate liquor** — finely ground roasted cacao bean nibs, i.e., pure chocolate without any additives; also called bitter chocolate and bakers' chocolate.

**chocolate products** — foods made from finely ground cacao beans mixed with other ingredients such as sugar, milk, etc. Some of the standard forms are bitter chocolate, milk chocolate, and sweet chocolate.

**Choco-roles** — *Mexico* a trade name for "Swiss rolls" containing pineapple jelly filling and covered with a chocolate-flavored coating.

**choklad** — *Sweden* chocolate.

**cholesterol** — an important physiologically active fat-soluble compound supposedly found only in every animal cell and in no plant cells.

**chop** — (n) the product of a break operation in a roller mill.

**chopin** — *Poland* two shortbread cookies with jam deposited between them.

**Chorleywood bread process** — a bread dough processing method that uses high speed development in special mixers to replace the dough conditioning that normally occurs during bulk fermentation. Also relies on heavy supplementation of the dough with several chemical modifiers.

**choux** — also, "pâte à choux." A type of vapor-leavened dough or batter used for making eclair and cream puff cases. Produced by beating eggs with a mixture of fat and gelatinized (with hot water) flour. When properly baked, large bubbles form inside the piece and then collapse so as to give an almost empty interior.

**Christmas pudding** — *UK* a dough of flour, water, suet, and dried fruit that has been cooked by steaming. Served warm topped with brandy butter or custard.

**chrust chyli faworki** — *Poland* said to mean "kindling" or "favors," a kind of fried cookie typically made of flour, butter, sugar, egg, vinegar, and sour cream; not leavened; an Easter dessert.

**chrusciki** — *Poland* a fried cookie, unleavened or with a small amount of baking soda. Sometimes flavored with whisky. The cooked pieces are usually coated with powdered sugar.

**chunk** — a grain particle composed mainly of bran, but having some endosperm attached. These particles are produced when milling conditions are too extreme.

**churyek** — *Georgia* unleavened bread with sesame seeds.

**churro** — *Spain* a more or less sweet fried dough usually in stick form (often with longitudinal corrugations) but sometimes in doughnut shape.

**ciabotta** — *Italy* lean formula sourdough hearth-bread in a loaf-shape somewhat similar to the traditional Italian loaf, but with an irregular slump in the center third, the shape said to be reminiscent of a "slipper."

**ciambella** — *Italy* ringshaped bun.

**ciasto kefirowe** — *Poland* a chemically leavened cake made with kefir as the liquid ingredient; buttermilk is often substituted for kefir.

**cimet** — *Serbia/Croatia* cinnamon.

**cinnamon** — a flavoring material made from the bark of the evergreen tree, *Cinnamomum zeylanicum*, Ceylon cinnamon. Cassia is often called "cinnamon," and both spices are used in the same types of foods. A very popular flavoring for sweet doughs. Commercially, offered as "quills" (thin dried bark rolled up in cylinders), powdered bark, extracts, oils, etc. The warm, aromatic flavor is highly accepted by most consumers.

**cioccolata** — *Italy* chocolate.

**C.I.P. systems** — clean-in-place systems include vats in which cleaning solutions are prepared, high pressure pumps for transferring the solutions, and spray balls inside the food processing unit to disperse the solution to all surfaces requiring treatment. Draining and collecting means are also required for the effluent.

**ciruela** — *Spain* plum.

**ciruela pasa** — *Spain* prune.

**ciste** — *Ireland* crust for meat pies, typically made with suet, milk, and self-rising flour, and sometimes containing raisins.

**citric acid** — an organic acid found in many fruits, such as oranges and lemons. Most commercial citric acid has been prepared by oxidation of glucose. Large quantities of this material are used in the beverage industry, but it is less important in bakery formulations.

**citroen** — *Netherlands* lemon.

**citron** — a fruit of the citrus family now used only as a source of rind for candying; often a component of fruit cakes.

**citron** — *France Sweden Denmark* lemon.

**clafouti** — *France* a dessert prepared by baking a rich, light egg batter poured over a layer of berries or other fruit deposited in (usually) a round fluted pan.

**clarify** — (1) To prepare drawn butter by melting butter, allowing the sediment and water to collect at the bottom, then pouring off the clear supernatant, which consists almost entirely of butterfat. (2) To remove insoluble materials from fruit juices, etc., by filtration, centrifugation, or settling.

**clé** — *France* the outer seam in a dough strip that has been molded into a loaf.

**cleaning, of grain** — the removal from a lot of wheat, or other cereal kernels, by special machines, of foreign seeds, broken grain, shriveled seeds, miscellaneous debris, etc..

**clear flour** — the portion of flour collected in a mill after the "patent" mill streams have been diverted.

**clearing time** — time from beginning of mixing until the dough forms into a single mass and takes up the material smeared on the walls of the bowl.

**clears** — the coarser parts of a straight flour. These mill streams are sometimes divided into first clears, second clears, etc.

**close-textured** — describes the interior of a loaf or roll that has uniform and small vesicles.

**cloverleaf roll** — a roll of about two ounces weight formed from three balls of oiled bread roll dough placed in a muffin cup; when baked, the roll exhibits three rounded lobes at the top and is easily separated into three pieces. Also prepared in automated molding systems by pressing a single dough ball with a Y-shaped cutter.

**cloves** — a spice of penetrating aroma and hot pungent taste, made by drying the unopened flower buds of a tropical tree, *Eugenia caryophyllata*. It is used fairly widely in bakery products, probably mostly in mince pie fillings, but it is also compatible with pumpkin, cherry, and apple fillings.

**club wheat** — variety of *Triticum æstivum*, not a commercial crop in the US.

**coaters** or, "coating machines." Equipment that applies seasoning or a flavored glaze to, e.g., popped corn, often by tumbling the food particles in the presence of a spray of the coating mixture.

**cobbler** — a baked dessert something like a pie, but generally baked in a deeper pan and sometimes having only the top crust. Some versions have the top crust formed of a sweetened soda biscuit dough.

**cob-cured popcorn** — harvested cobs of popcorn that are allowed to equilibrate or temper, typically in a bin, before shelling.

**cocada** — *Portugal* coconut macaroon.

**cocket** — *Old English* medieval term for dark bread that is lighter than whole wheat or rye bread.

**cocoa** — a powder made from chocolate from which most of the fat has been pressed or extracted.

**cocoabeans** — the seed of the cocoa tree, after fermenting and drying, consisting of a fairly thin shell surrounding the fairly hard contents that will become "nibs" when roasted and cracked.

## GLOSSARY OF CEREAL TECHNOLOGY TERMS

**cocoabutter** — fat pressed from ground (and usually roasted) cacao nibs.

**cocoa liquor** — finely ground cocoa nibs, after they have been roasted and, sometimes, dutched. Essentially synonymous with bitter chocolate, bakers' chocolate, baking chocolate, etc.

**cocoa nibs** — the interior of a cocoa bean after it has been heat-treated then broken into bits of varying size; this material is an intermediates in chocolate processing and will be further heat treated, and sometimes alkalized, then finely ground to form cocoa liquor. Not used directly as a food ingredient.

**coconut** — the fruit of *Cocus nucifera*, the coconut palm, the edible flesh of whihc is available commercially in many forms.

**coconut cream** — thick coconut milk.

**coconut milk** — made by grating coconut meat and compressing it, the thick fluid usually being thinned with water.

**coconut meat** — the thick layer of firm white material found inside the shell of a matured coconut; high in oil content. Dried, shredded, grated, toasted, sweetened, and subjected to many other operations in forming commercially offered food ingredients. Also known as copra when dried.

**coconut oil** — a highly saturated fat pressed from the dried meat of the coconut. It is a bland white fat having a melting point above that of some other common ingredient oils, such as soybean or cottonseed oils. It is very useful in icings, etc. Commercial variations include various fractions of the oil that have been separated so as to have different melting points.

**coconut, sweetened** — also, "prepared coconut." Coconut shreds or granules that have been sweetened and softened by mixing with, e.g., glycerine and various sweeteners such as corn syrup. Used as toppings on confections and baked products, and as ingredients for cakes, cookies, pie fillings, candies, etc.

**coconut water** — the dilute, slightly opaque liquid found in a fresh coconut. Seldom used as an ingredient in the West.

**Codex Alimentarius** — an "official" collection of ingredient and food product standards and specifications prepared by an international organization.

**coentro** — *Portugal* coriander.

**coffee cake** — a multi-serving size of baked sweet yeast-leavened dough made in various shapes and, usually, with fillings or toppings. Dough formulas vary from fairly lean to very rich. Both chemically leavened and fermented dough varieties are known.

**colchones** — *Mexico* orange-flavored sweet bread.

**cold test** — a test that determines the extent to which the high melting-point fractions of an oil have been removed during the production of "winterized" oil; the oil is held in an ice water bath and the time required for the first appearance of cloudiness is recorded as "Cold Test Hours."

**coliforms** — gram negative bacteria that can ferment lactose. The most familiar member of this large group is *Escherichia coli*, found in vast quantities in the gut of man and many other animals.

**color additive** — any material, natural or synthetic, that is applied to (or used as an ingredient in) a foodstuff for the purpose of changing the hue of the latter.

**coloration** — act or art of coloring; state of being colored.

**combination bakery** — a bakery that uses more than one form of production or supply, e.g., a bakery that uses both refrigerated dough bake-off and scratch-mix production methods.

**combination icings** — blends of two or more of the basic types of icings, e.g., a mixture of marshmallow icing and cream icing.

**combine** — a grain harvesting machine combined with a threshing device, so that the wheat or like plants can be headed, threshed, and cleaned in one pass through the field.

**cominho** — *Portugal* caraway.

**compact oven** — a smaller version of the revolving tray oven in which the trays revolve from side to side, instead of being carried from front to back. This allows placement of the oven in a shallow space that is more convenient for many food service operators and bake-off boutiques.

**composite flour** — wheat flour blended with some nonwheat meal or flour (such as rye flour).

**compound** — (1) A chemically defined substance consisting of two or more elements combined in fixed proportions. (2) A prepared mixture of some kind.

**compound coatings** — imitation chocolates made by combining (for example) cocoa, coconut oil, and sugar. Other flavors besides chocolate can also be made in this way. The essential difference between true chocolates and compound coatings is that little or no cocoa butter is included in the latter.

**compound shortening** — a mixture of animal and vegetable fats processed as shortening.

**compressed yeast** — this ingredient consists of undried cells of living yeast that have been combined with fillers and pressed into cakes; it must be distributed and stored under refrigerated conditions.

**compression board** — in a molder, a strip or plate that applies pressure to the curled piece of dough as it is carried beneath the strip or plate; can be adjusted to vary the force applied to the dough piece.

**compression chamber** — (1) In a dough divider, the cavity of defined dimensions into which dough is forced to measure the desired portion size. (2) In a dough molder, the section where pressure is applied to the curled piece of dough so as to seal the layers together.

**compressor** — the pump in a refrigerating unit; it raises the pressure of the refrigerant before it passes to the condenser.

**concentrated milk products** — a class of partially dried milk products, in the form of thick liquids or plastic materials, such as condensed whole milk or concentrated sweetened skim milk.

**conchas** — *Mexico* rounded pieces of pan fino made up in the shape of a sea shell and sometimes topped with sweet paste.

**conching** — a processing step applied to premium chocolates, consisting of a prolonged working or grinding at elevated temperatures for the purpose of improving the finished product's "smoothness" and mellowing or blending its flavor.

**condensation** — in general, the transition of a substance from the vapor phase to a liquid state. Also, the formation of water droplets ("dew") on the surface of a relatively cold solid surface that has been exposed to a warmer atmosphere.

**condensed milk** — whole fluid milk from which a substantial portion of the water content has been removed by evaporation. Usually contains a high percentage of added sugar.

**condenser** — the unit in a refrigeration system that receives the hot, high pressure refrigerant gas from the compressor and cools it until the gas returns to the liquid state.

**condiment** — a material that can be added to a prepared food to give it added zest or appeal; mustard, ketchup, and pepper are examples.

**conditioning** — (1) In grain milling, controlled moistening of grain to prepare it for grinding; the bran is toughened and the endosperm softened by the added moisture. Heating is involved in some of these operations. Similar to tempering. (2) In baking processes, adjusting mixing parameters, rest periods, etc., so as to yield a dough that will respond to machining without problems and will result in a finished product of good quality.

**conduction** — a method of heating or cooling in which the basic physical mechanism is the transfer of energy from one solid to a contacting solid material.

**confectioners' coatings** — also, "compound coatings." Enrobing materials resembling chocolate coatings in their physical properties and eating texture, but sometimes white or pastel-colored and variously flavored. Generally contain no cocoabutter.

**confectioners' sugar** — cane or beet sugar that has been finely ground and sifted so as to yield a smooth powder suitable for use in icings, buttercream fillings, and the like; it is commercially available in two or more particle size distributions, and normally contains about 3% cornstarch to prevent caking. Synonymous with "powdered sugar" and "icing sugar."

**confectionery** — a category or class of prepared sweet dessert-type foods including all kinds of candy and some bakery products.

**confectionery fats** — a broad range of ingredients used in the formation of sweet confections, including some bakery products. Their primary application is in the formulation of compound coatings, both chocolate-substitutes and white and colored coatings. Often based on fractionated coconut oil.

**confiture** — *France* jam.

**congeal** — to change from a liquid into a solid (or semi-solid) mass.

**congee** — *China* a rice gruel, porridge, or soup of thin consistency, to which is added various condiments. Versions made from millet are also known, esp. in North China.

**conjugation** — a type of unsaturation in fatty acids; a repetition in one fatty acid molecule of a sequence of alternating saturated and unsaturated bonded carbon atoms.

**consistency** — as applied to dough, means the tactile evaluation or "feel' of the dough. It is one of the subjective criteria by which the proper absorption is judged.

**continuous breadmaking systems** — an assemblage of equipment that converts bulk ingredients into a finished dough in an uninterrupted sequence of automated operations.

**continuous developers** — essentially a kind of mixer that accepts a continuous input of mixed but incompletely developed dough that is in a relatively soft, plastic, and inelastic form, and by input of oriented force from paddles, augers, or the like converts the received material into an elastic, expandable dough suitable for depositing into baking pans.

**continuous fermenters, positive flow** — a brew- or liquid sponge-fermenting system in which a pumpable premix is continuously delivered to the fermenting tanks at a rate equivalent to the rate at which the finished brew is drawn off for use in the dough.

**continuous mixing** — any kind of mixing process that converts ingredients into doughs or batters without the necessity for separating the output into batches.

**contre-frase** — *France* making the dough tougher by adding flour to it during kneading.

**control** — (n) a sample or experimental specimen that represents the standard material before any of the variations being tested have been applied.

**controlled atmosphere packaging** — a method of extending the shelf-life of foods by reducing or eliminating the amount of oxygen in the package and, sometimes, by adding gases such as carbon dioxide.

**convection** — heat transmission by moving currents of gases or liquids.

**convection ovens** — any oven in which the main heat transfer method is moving air. The addition of fans and blowers provide forced convection that speeds up baking.

**conversion** — in brewing, a holding period during which the mash is typically maintained at 162°F to 167°F for about 10 to 20 minutes, providing optimum conditions for alpha-amylase activity.

**conveyor** — any type of fixed-in-place device that can continuously transfer material from one location to another; rollers, belts, augers, buckets on chains, and air flowing through pipes are some of the common moving forces and mechanisms used in conveyors.

**conveyor ovens** — baking systems utilizing a chain-type of conveyor that carries pans or straps in a convoluted path through a heated chamber, nearly always combined with a conveyorized proofing system.

**cookie** — a sweet baked product of small size, typically containing flour, sugar, shortening, flavoring, and other ingredients; relatively low in moisture content. Cookies are seldom yeast leavened, but all other types of leavening systems have been used. Generally, they are relatively dense as compared to cakes. Characteristics of examples of this class vary widely.

**cookie bag** — a canvas or plastic bag of roughly conical form that can be filled with dough and squeezed by hand to force the dough through a metal or plastic orifice at the tip of the cone.

**cookie crusts** — although there are some examples of pie crusts prepared from cookie doughs formed in a pie pan of conventional shape, then baked, the usual cookie crust is made by pressure-forming a dry mixture of granulated baked cookies, shortening, and sugar.

**cookie cutter** — any type of hand-operated device that can be used to cut shaped pieces from a sheet of cookie dough.

**cookie sheet** — a flat metal sheet, usually without elevated rims, that can be used to support cookie pieces while they are being transferred into the oven, and to hold them during baking.

**cooler** — (1) A refrigerated chamber held above freezing temperatures. (2) A conveyor or chamber used for bringing hot loaves and rolls to about room temperature, usually by exposing them to currents of ambient air.

**cooling conveyor** — usually a system of belts suspended from the ceiling and equipped with a fan system that draws in air at the discharge end and moves it over hot loaves that have just emerged from the oven.

**copata** — *Italy* small wafer made principally of honey and nuts.

**corbeille** — *France* a small vacherin.

**core** — in describing baked product texture, this refers to condensed or solid regions within the crumb that appear to have not undergone any significant expansion; it is a serious quality defect.

**coriander** — the leaves and seeds of *Coriandrum sativum*, used as flavorants. The dried, globular, brownish seeds have a slight fragrance and a pleasant taste. Although coriander has been used in many kinds of bakery products, it is not one of the favorite spices for this purpose in the U.S.

**coriolus mass flow meter** — an electronic device that measures the deflection of a vibrating pipe through which a liquid is flowing to give an indication of the flow rate; sometimes used as elements in a continuous measuring system.

**corn** — the plant *Zea mays* and its seed; it exists in various types such as sweet corn, field corn, and popcorn. In the U.K., the term "corn" may be applied to almost any kind of grain.

**corn bran** — the fibrous outer coating of the corn kernel, regarded as a low-grade food for cattle or a high-grade food for humans.

**corn bread** — a multi-serving cake made from a chemically leavened dough or batter containing variable proportions of wheat flour and corn meal (almost always yellow corn meal), slightly to moderately sweetened. Used as a bread. Identical and similar batters are made into corn muffins.

**corn chips** — thin, crisp, pieces of cornmeal dough or masa that have been fried or baked, then (usually) seasoned with salt, spices, and other flavoring materials. Shape is often triangular, but round and irregular chips are also seen.

**corn cones** — ground corn of particle size intermediate between meal and flour, used in the bakery for dusting dough pieces and pans.

**corn curls** — a snack product consisting of extrusion puffed cornmeal, flavored and colored, in the form of short, irregular cylinders.

**cornet** — *France* a cone-shaped pastry. *UK* cone for ice cream.

**cornetti** — *Italy* crescent rolls.

**corn flakes** — thin, crisp, irregularly-shaped pieces formed by dry heating of pellets or chunks of corn that have been gelatinized by wet cooking then passed between rotating steel cylinders to reduce their thickness. Malt, corn syrup, salt, and other seasonings are added prior to the flaking step.

**corn flour** — a very fine granular form of maize endosperm; in the UK, the term is sometimes applied to corn starch.

**corn gluten** — in the corn wet-milling process, the dried aqueous fraction remaining after the deposition and removal of starch from slurried corn endosperm; it is relatively high in protein content and is used mainly as a constituent of animal feed.

**Cornish pasty** — *UK* a fairly large individual baked pie of the turnover variety, filled with chopped meat, potatoes, onions, etc.

**Cornish split** — *UK* a plain sweet roll ("bun") cut in half and filled with preserves and cream.

**cornmeal** — meal of various granulations produced by dry-milling kernels of white or yellow field corn, usually with the germ removed to delay development of rancidity.

**cornmeal boards** — plastic, fiberglass, or wooden sheets used to support molded dough pieces during transfer and proofing, so called because cornmeal is dusted on the surface of the sheet to prevent adhesion of the dough. Also called proofing boards.

**corn muffins** — muffins of cupcake size and shape made from the same slightly sweet and chemically leavened type of doughs and batters used for cornbread.

## GLOSSARY OF CEREAL TECHNOLOGY TERMS 45

**corn oil** — the cooking and ingredient oil that is the purified fatty material pressed or extracted from corn embryos removed in the corn wet-milling process.

**corn starch** — a fine white powder made from corn kernels by the wet milling process.

**corn sugar** — dried glucose made from corn syrups.

**corn syrup** — viscous liquids containing mixtures of sugars and other carbohydrates and made by the controlled hydrolysis of corn starch; many different kinds are available — they vary in the kind and amounts of carbohydrates they contain, the total concentration of solids, etc.

**corn syrup solids** — corn syrup that has been dried and then ground into a fine powder.

**cortadillo** — *Spain* small pancake with lemon.

**cottage loaf** — a round loaf of bread, formed by placing a relatively small ball of dough on a larger ball of the same type of dough just before proofing, the conditions being adjusted so the two pieces of dough merge at the contact surface, but otherwise retain their identity.

**cotto** — *Italy* cooked.

**cottonseed flour** — milled cottonseed cake that results from the oil pressing operation. Usually heat treated to deactivate certain toxic factors (gossypol) in the meal.

**cottonseed oil** — the edible refined oil pressed or extracted from cottonseed. Highly regarded as a constituent of hydrogenated shortenings.

**coulibiac** — see "koulibiac."

**coup** — *France* dish containing a single scoop of ice cream which is usually decorated with syrups, fruit pieces, and whipped cream.

**coupe-pâte** — *France* bowl scraper or dough knife.

**couper le pâton** — *France* to slit the dough just before it enters the oven.

**couronne** — *France* crown-shaped loaf, formed by baking the dough in a basket having a center tube.

**couscous** — small basically round pellets of various size made by agitating semolina (from durum or millet) in the presence of hot water vapor.

**couverture** — enrobing materials of which chocolate is the prototype; pastel-colored fatty coatings of non-cacao origin are also placed in this category.

**crackling bread** — also, "cracklin bread." Corn bread containing bits of fried pork rind (or the residue from lard rendering) or bacon.

**cracknel** — a plain cracker made originally of only white flour, eggs, and sugar (no water); various shapes. The best examples have an unusually low density and a very fine grain, and exhibit a smooth and shiny surface.

**crack stage** — in sugar boiling, the condition reached by a syrup brought to 280°F. If a drop or string of the sugar is cooled in cold water, it cracks when deformed.

**cream** — (1) A fluid milk product enriched in fat to different levels, examples being half-and-half and whipping cream. (2) Also, "creme" or "crème," A thickened cooked plastic mass typically containing sugar, an egg ingredient, milk, and a viscosity improver (starch, gelatin, etc.) in addition to the usual flavorings, colorings, and adjuncts — used for filling pies, doughnuts, etc. (3) A flowable pudding, especially if whipped.

**cream horn** — a hollow cylinder formed from a spirally wound strip of puff pastry, to be filled after baking with whipped cream or the like.

**creaming** — the process of mixing and aerating a shortening and one or more dry ingredients, such as sugar or flour.

**cream injector** — a device for extruding cream filling from a reservoir through a tube into the centers of baked or fried dough pieces such as doughnuts and snack cakes.

**Cream of Rice** — trade name for a granular form of rice endosperm, used as a cooked breakfast cereal.

**Cream of Rye** — trade name for a granular form of rye endosperm, used as a cooked breakfast cereal.

**cream of tartar** — acid potassium tartrate=potassiium bitartrate; one of the acid-reacting substances used in baking powders. In the ingredient form, it is a white powder. It is also used in small amounts to facilitate the whipping of egg whites and as a "doctor" or sugar hydrolyzing agent in the boiling of sucrose syrups.

**Cream of Wheat** — trade name for a granular form of wheat endosperm, used as a cooked breakfast cereal.

**cream pies** — pudding or custard type fillings poured into pre-baked pie crusts; usually topped with whipped cream.

**cream puffs** — hollow balls of baked cream puff dough (chou paste) that have been filled with starch-based pudding, cooked custard, or whipped cream.

**cream yeast** — or, "liguid cream yeast." A commercial bakers' yeast preparation having a moisture level of 78% to 80%; it can be handled by liquid transfer equipment and is immediately active when mixed with doughs. Must be stored at 38°-40°F.

**crease** — the part of the outer covering of a wheat kernel that constitutes the inward folding of the bran layers to make a relatively deep groove extending the length of the kernel on one side.

**crema** — *Spain Italy* cream.

**crema batida** — *Spain* whipped cream.

**creme** — *Portugal* cream.

**crème chantilly** — dairy cream whipped with vanilla and sugar; used as a filling or topping for sweet baked products.

**crème de riz** — *France* fine rice flour, used for thickening sauces.

**crème fouettée** — *France* whipped cream.

**crepe** — or, "crêpe." Very thin pancake of the French type; there are no added leaveners in the traditional recipe but commercial versions may contain some baking powder.

**crepe soufflé normandie** — a thick sweet crepe or a thin cake baked with a layer of sauteed apples between two crepes, then topped with the apple pan juice when served. The batter is air-leaved with beaten egg whites.

**crepe suzette** — a sweetened crepe flavored with lemon rind/oil and orange liqueur; after baking, flamed in a sauce consisting of butter, sugar, orange juice, and various liqueurs. Other ingredients may be added.

**crescent roll** — a yeast-leavened bread roll made by rolling up a triangle of sheeted roll-in dough so that the center is much thicker than the ends, then curving the dough piece into a crescent shape.

**cripple** — a damaged or otherwise defective and unusable baked product.

**crisp** — (n) (1) A dessert of sweetened cooked fruit covered with a mixture of oats and brown sugar, then baked. (2) A thin, relatively large disc of yeast-leavened sweet dough that has thickly coated (on one or both sides) with a mixture of granulated sugar and cinnamon before baking to a fairly low moisture content.

**crisps** — *UK* potato chips.

**crni hleb** — *Serbia/Croatia* black bread.

**croissant** — now often used as a synonym for crescent roll, but formerly used only for similarly shaped pieces made with a rich but not sweet roll-in dough. Yeast-leavened.

**croquembouche** — small cream puffs stuck together in the form of a pyramid or cone, using caramelized sugar as the adhesive. Often, elaborately decorated.

**cross-grain molder** — a loaf molder that takes the sheeted dough piece and transfers it to a forming conveyor placed at a right angle to the flow of the sheeting conveyor. Its purpose is to roll up the dough perpendicularly to the direction of sheeting, so as to improve the uniformity of moisture distribution in the finished dough piece.

**cross panning** — methods that attempt to achieve the same or better results as in cross-grain molding, without using the cross-grain molder.

**crostada** — *Italy* pie, flan.

**crostini** — *Italy* croutons.

**croustade** — *France* shell made from pie dough or puff pastry, baked or deep-fried and then filled with fish, seafood, meat, or vegetables; also can be made from a hollowed-out roll of baked yeast-leavened bread.

**croûte** — *France* either a crust or a disc or finger of toasted or fried bread. Also, *en croûte* means baked in a covering or casing.

**croutons** — or croûtons. In present day commercial bakery usage, "croutons" is generally understood to mean cubed or crumbled dried bread that has been seasoned with herbs and spices and sprayed with oil to serve as an additive to soups or as a basis for poultry stuffing.

**crown** — inside top of the baking chamber of an oven, especially applied to large stationary hearth ovens.

**crullers** — fried dough products usually in the form of long (6 to 8 in) double twists, though sometimes in a fancy doughnut (circular) shape.

**crumb** — in leavened baked goods with a crust, all of the dough product except the crust.

**crumble** — (n) sweetened and otherwise flavored fruit filling topped with a fairly thick layer of streusel and baked.

**crumbs** — small fragments formed in any manner from larger pieces of baked products.

**crumpet** — *UK* there are many points of similarity between crumpets and English muffins; Both are made from thin yeast-leavened doughs or thick batters, and are generally cooked on a griddle, often in restraining rings. Both are usually toasted immediately before consumption. Most commercial versions of crumpets appear to be moister and more gelatinous in texture than muffins, they have many more large holes (bubbles) on their surface; and, their crust is not as clearly differentiated from the crumb.

**crust** — the distinctive outside layer of a baked product. Includes all of the portions that have undergone browning or substantial dehydration and consolidation during baking. By extension, a hard or crisp layer on any normally soft piece of food.

**crusting** — the formation of dry surface layers on dough pieces due either to loss of moisture or accumulation of dusting flour. Generally considered very undesirable.

**cryogenic freezing** — freezing methods that use liquefied carbon dioxide or nitrogen to withdraw heat from products.

**crystallization** — condensation from solutions or molten masses of particles which exhibit uniform arrangements on the atomic or molecular scale and that are nearly pure.

**crystallized** — (1) A solid material that has been deposited in the form of more or less uniformly shaped (but different size) granules, from saturated solutions or molten masses; not ground. In the food industry, refers particularly to ingredients such as sugar or salt particles that are in recognizable crystalline forms. (2) Fruit pieces and the like that have been steeped in concentrated sugar solutions until their tissues have been completely saturated with sugar and, finally, rolled in or sprinkled with sugar crystals; these materials are used as confections and for decorating and otherwise enhancing bakery foods and confections.

**cube sugar** — pure cane or beet sugar that has been crystallized in blocks or sheets and then cut into cubes. Other methods include crystallizing the sugar directly into cube molds.

**cuernitos** — *Mexico* relatively small cuernos.

**cuernos** — *Mexico* crescent shapes of pan fino, with flavor paste applied to the outside before baking.
**cumin** — similar in appearance to caraway seeds, cumin seeds do not have the licorice flavor of the former spice. Used in curries, and occasionally in snack foods.
**cup** — a standard measuring cup, as referred to in many consumer recipes, contains 8 fluid ounces. The actual volumes contained by other kinds of cups cover a wide range.
**cup cakes** — small sweet muffins baked from cake-type batter deposited in muffin pans, often iced and decorated, sometimes filled.
**curacao** — an alcoholic liqueur that has been flavored with dried orange peel; occasionally used for flavoring gourmet icings, fillings, etc.
**curd** — the casein lumps that form when milk is coagulated with rennet and/or acids.
**curdled** — batters that appear to have separated into liquid and pasty fractions, somewhat resembling curdled milk.
**curlers** — also, "curling rollers." Devices that roll up a sheet of dough, with or without adjuncts, and form it into a cylinder. Curlers are essential parts of the molding process for bread loaves made in a traditional plant.
**curling chain** — in a bread molder, a chain belt that pulls up the leading edge of a dough sheet and begins the curling action that eventually forms a cylinder.
**curling cylinder** — or, curling roller. A metal cylinder, usually power-rotated, that is positioned diagonally across a fabric or plastic belt upon which a sheet of baked or unbaked pastry (often with filling paste applied) is being conveyed. The function of the cylinder is to form a continuous roll, rope, or cylinder having a more or less spiral cross-section. They are used in one of the steps of mass producing cinnamon rolls and the like.
**currants** — the acidulous berry of the currant bush, somewhat similar to gooseberries, but in the baking trade "currants" means small raisins (i.e., dried grapes).
**custard** — basically, a sweetened mixture of eggs and milk that has been cooked over hot water. Hundreds of recipes exist, some with neither eggs nor milk.
**cut-in** — to mix solid fat and flour manually or with a pastry blender or machine so that the shortening is distributed more or less uniformly throughout the mixture in the form of flour-covered but separate lumps.
**cutlin pudding** — *Ireland* a boiled pudding similar to plum pudding, but usually somewhat simpler in composition, containing fewer kinds of dried fruit and less spicing.
**cut-off** — adjustable dividing board located under a sieve to enable the miller to change the flow of stock; purpose: to be able to adjust the separation as grinding and sifting conditions change.

**cutting dies** — metal or plastic sheets or strips formed into designs that, when pressed into sheets of dough, cut out pieces of desired shape.

**cutting machines** — forming apparatuses for cookies and crackers that cut pieces of finished shape from a sheet of dough having uniform thickness.

**cutting roller** — the more rapidly rotating member of a pair of cylinders in a flour mill roll-stand.

**cvibak** — *Serbia/Croatia* rusks.

**cwibak** — *Poland* cake, especially fruitcake.

**cysteine** — a sulfur-containing amino acid. It is used as a reducing agent to modify the physical properties of doughs.

## -D-

**dadel** — *Netherlands* date.
**dadler** — *Denmark* dates.
**dairy blends** — dry mixtures intended to replace nonfat dry milk or other dairy ingredients in bakery products; typically, they consist of dried whey mixed with soy protein concentrates.
**dalchini** — *India* cinnamon.
**damasco** — *Mexico* apricot.
**damper** — an adjustable plate or other device placed in a duct to permit adjustments in the flow of smoke or air to or from a combustion chamber.
**dampet brunbrod** — *Denmark* steamed brown bread, chemically leavened, usually containing raisins.
**dampfnudel** — *Germany* steamed, sweet dumpling, usually served warm with vanilla sauce.
**Danish pastry** — originally, a very rich, flaky yeast dough that had been laminated with butter or other shortening. Now, often refers to any kind of sweet breakfast pastry, even those not made from roll-in dough.
**dark meal** — in cookie bakery parlance, means the ground-up scrap of dark-colored and strong-flavored cookies that have been rejected (or returned) for some reason. Used as a bulking or non-characterizing ingredient in dark-colored doughs such as molasses cookie formulas.
**date filling** — ground or chopped dates mixed with water and other ingredients and cooked.
**DATEM** — diacetyl tartaric acid esters of monoglycerides; a class of chemical compounds that affect the properties of dough; they are employed mainly to increase loaf volume and soften the crumb.
**date paste** — pitted dates ground to a small particle size, yielding a homogeneous dark-colored, very stiff mass.
**dater** — an apparatus used to imprint a date code on a package.
**dates** — the fruit of a species of palm, *Phoenix dactylifera*; there are several varieties, all very high in sugar and low in acidity. Available dried, with and without pits, and ground as pastes.°
**dátil** — *Spain* date.
**dattero** — *Italy* date.
**datule** — *Serbia/Croatia* dates.
**dau phong rang** — *Vietnam* roasted or fried peanuts, usually unblanched.
**deck oven** — small oven with stationary hearth, often heated by electricity. Usually loaded and unloaded with peels. Widely used in pizza shops.
**decoction mashing** — a method of preparing wort (primarily for bottom fermentation beers) in which part of the mash is withdrawn, boiled, and then returned to the main tun to raise the temperature of its contents.

**decorating tubes** — metal or plastic orifices to be placed in the opening at the point of pastry bags so as to form the extruded icing into fancy shapes.

**decoration** — something added to a food to improve its visual esthetics and which does not have as its primary purpose improving the flavor, texture, nutritional value, etc., of the edible object.

**defatted soy flour** — soybean meal from which substantially all the fat has been removed; it is almost always heat-treated to inactivate undesirable components and improve flavor. Used as an inexpensive protein supplement and to bind water in doughs and batters.

**defect** — failure of some quality factor of a product (or ingredient or package) to reach an acceptable level.

**degasser** — a machine or device that works the dough prior to dividing so that much of the gas can escape; usually it is a kind of pump or kneader placed at or near the divider hopper.

**degermination** — the process of removing the plant embryo (or germ) by mechanical devices, a term restricted almost exclusively to corn and rice milling.

**degree** — a unit on a scale of measurement such as Fahrenheit for temperature, Brix for sugar concentration, etc.

**dehumidification** — removal of moisture vapor from the atmosphere.

**dehydrate** — to remove water from a substance by any method, although not usually applied to procedures that strain or press liquid water from a mass.

**deionized water** — water from which some of the ions have been removed by passing the liquid through beds of ion-exchange resins, a common water-softening procedure; often said to be "equivalent" to distilled water, but it is not.

**delidder** — a machine for automatically removing the covers from pans of sandwich bread and the like.

**demarara sugar** — *UK* brown sugar, usually of the semi-refined type.

**demi-glacè** — *France* soft frozen ice cream.

**dendritic salt** — salt that has been crystallized in the presence of certain additives so that it forms branched crystals; said to be more adherent than ordinary salt to potato chips and the like.

**dense phase transfer** — a pneumatic conveying method involving relatively low air-to-solids ratios.

**density** — mass per unit volume.

**dent corn** — the most common type of field corn; it is characterised by kernels which have an indentation at the top of the broad end.

**deodorization** — the last step in traditional processing of edible fats and oils. It usually removes the relatively volatile trace components (such as aldehydes, ketones, alcohols, and free fatty acids) that contribute undesirable colors, odors, and flavors. Typically, deodorization gives an oil that has less than 0.05% free fatty acid content and that is nearly tasteless and odorless.

**depanners, automatic** — machines that remove baked products (such as bread loaves) from pans and, usually, convey the food to packaging equipment and the pans to a temporary storage area.

**deposit cookies** — cookies made from very soft doughs that are extruded directly onto the oven band and, usually, not cut off at the die orifice by a knife or wire .

**depositer** — a scaling device that extrudes or drops a measured amount of material (such as muffin batter) into a pan or other receptacle.

**derivative** — a substance obtained by chemically combining one chemical compound with another.

**desalinization** — the partial or total removal of salt from water, so as to make sea water or brackish water potable or at least usable for some agricultural or commercial purpose.

**desiccant** — in the lexicon of food processing and packaging, a "desiccant" usually means a solid material (such as silica gel) having a very low water activity that can be used to take up moisture vapor from the gases in contact with food particles.

**desiccant pouch** — also, "desiccant sachet." A porous packet or pouch containing desiccant particles; it can be placed in containers of low moisture food proucts (e.g., potato chips, popcorn) to prevent the uptake of water vapor by the food.

**desiccate** — to dry by removal of water in the vapor state.

**design printers** — any device, based on whatever principle, that deposits or imprints material forming a decorative image on an object, particularly applied to machines that decorate cookies with a recognizable image.

**detailer rods** — or, "detailers." On a chocolate enrobing line, simple devices that remove drips and trailing strings ("tails") of chocolate from coated pieces; usually  metal rods of small diameter rotating counter to the movement of the food piece and placed so they contact the bottoms of the pieces.

**detergent** — a surface active agent, other than soap, that can be mixed with water to increase the effectiveness of the fluid in removing grease, soil, and dirt from a surface.

**developing** — the process of mixing a bread dough (or the like) for a time, at a speed, and under the conditions required to cause the dough to exhibit the best processing response of which it is capable. If the flour is of good quality and the formula balanced, a bread dough can be developed to a soft but very elastic and extensible mass that retains leavening gases and goes through make-up machinery without becoming weak and sticky. Developing involves both chemical and physical changes, some of them poorly understood.

**devil's food cake** — a chocolate cake that has the leavening system adjusted so that the batter and cake crumb have a markedly alkaline reaction, leading to the development of a so-called mahogany color of the crumb.

**dew point** — the temperature at which moisture from the air will condense on a surface; an indication of the relative humidity of the air.

**dextrin** — an industrially useful material made by treating starch with acid or enzymes to partially hydrolyze it; has a low flavor (practically no sweetness) but is useful as a bulking agent, viscosity adjusting ingredient, and adhesive.

**dextrinizing enzyme** — alpha-amylase.

**dextrinizing rest** — in the conventional brewing process, a period during which the mash is held at (typically) 136°F to 162°F for 15 to 45 minutes, to provide optimum conditions for the activity of alpha-amylase.

**dextrose** — another name for commercial grade D-glucose ("corn sugar").

**dextrose equivalent** also, "DE." — a measure of the content of reducing sugar in a syrup or powder; it is calculated as though all of it were dextrose and reported as a percentage of the total dry substance in the ingredient; one of the essential specifications for corn syrup.

**dhania** — *India* coriander.

**dhiples** — *Greece* puff pastry filled with ground walnuts, with syrup poured over it after it has been baked.

**diacetyl** — a compound with a buttery flavor, present in doughs as well as in butter. It can be procured from flavor supply houses as a partially purified substance.

**diastase** — an enzyme that can convert starches into dextrose and maltose; synonymous with "amylase," which is the preferred term.

**diastatic malt** — malt or malt syrup that retains a considerable amount of the original amylolytic power of the sprouted barley from which the ingredient has been prepared; the more severe the heat treatments applied during manufacture, the lower the diastatic power of the malt.

**diastatic power** — a numerical designation of the starch-hydrolyzing power of an amylase preparation, such as malt syrup, as determined by a standardized test.

**dicalcium phosphate dihydrate** — occasionally used as a leavening acid to cause carbon dioxide release at a rather late stage in the baking procedure. Never used as the sole leavening acid in a system.

**diætmad** — *Denmark* diet food.

**die cups** — the shaped orifices through which dough is extruded in some kinds of cookie machines.

**die cylinders** — large rotatable metal or plastic cylinders containing patterned depressions on the surface that act as molds for cookie doughs pressed into them.

**dieettiruoka** — *Finland* diet food.

**dielectric heating** — a method of baking or cooking by a high frequency electromagnetic field that generates heat by causing rapid movement of some of the molecules making up the product. Has been commercially used to further dehydrate cookies or snacks that have passed through an ordinary oven.

**dies** — shaped metal or plastic orifices or cavities that are used in cookie manufacure to give the desired form to doughs.

**dietary fiber** — a food constituent that passes through the human intestinal tract without being digested. Includes not only true fibers such as cellulose, but hydrocolloid materials such as pectin. There have been many technical definitions and suggested analytical procedures for "dietary fiber."

**dietary food** — in its most inclusive sense, any food suitable for persons who are controlling their diet for physiological reasons, including control of caloric intake.

**dietetic** — identifies a food that allegedly has some nutritional modification of interest to health-minded consumers. Not to be confused with "diabetic" or "reduced calorie," which are terms of much narrower significance.

**differential** — the ratio of the speeds of rotation of fast and slow rolls in a pair of flour-milling rolls and the like.

**differential scanning calorimetry** — a technique for the study of heat flow in substances undergoing changes of temperature or phase. Such heat flow may result from changes in physical state (as from a liquid to a solid) or from polymorphic crystal transformation.

**digestive biscuit** — *UK* an older type of health cracker.

**diglyceride** — a chemical combination of fatty acids and glycerol in the proportion of two fatty acids to one glycerol residue.

**dijetalno** — *Serbia/Croatia* low calorie.

**dill** — dill seeds are the dried fruits of a herb. This spice has a flavor characterized by pungent, woody, and methol notes. It has been used in specialty breads and rolls, and as a component of some dusting powders used on savory snack chips, often combined with sour cream flavors.

**dilute phase transfer** — in pneumatic conveying technology, a method involving low solids to air ratios.

**dimethylpolysiloxane** — an antifoam agent used to reduce processing problems that arise when cooking products that have a tendency to foam excessively.

**dim sum** — *China* Cantonese dumplings, which can be made in a wide variety of fillings; cooked by steaming in bamboo utensils. Generally served as hors d'oeuvres or snacks.

**dingle pie** — *Ireland* a small meat pie with crust made from butter, flour, and water, similar in form and cooking method to Cornish pasties.

**dinner rolls** — bread rolls of almost any kind, shape, and size.

**dipless** — *Greece* fried cookies, see "kserotigna."

**diplomate** — *France* a custard containing crystallized fruit inside a structure of sponge fingers soaked with liqueur. Similar in concept to the English trifle.

**direct fired ovens** — baking equipment in which combustion occurs within the oven chamber; efficient use of fuel, but allows smoke and vapor from the fire to come in contact with the dough, which is generally undesirable.

**disaccharide** — a sugar, such as sucrose or maltose, made up of two monosaccharides, such as glucose or fructose, joined by chemical bonds.

**disc separator** — a machine containing a set of upright revolving discs covered with small pockets designed to fit (or reject) expected impurities, used in grain cleaning to remove foreign matter that has either a size or shape different from that of the desired kernels.

**disc slicers** — discs with horizontal rotating blades, used primarily for slicing partway through hamburger buns and hot dog rolls.

**disinfectant** — a chemical that can be applied to equipment or other surfaces to reduce the number of microorganisms present.

**dispenser** — any device that accomplishes the metered transfer of a material from a bulk supply to a finished or in-process food.

**disperse** — to distribute particles of a substance more or less uniformly throughout the total mass of another substance, as when spices are dispersed in a premix or nut granules are dispersed in an icing.

**displacement meters** — measuring devices for fluids that depend upon the repeated filling and emptying of a cavity of fixed size as their principle of operation.

**dissolve** — to cause a solid material to separate into its constituent molecules (or ions) and distribute uniformly through a liquid, as when sugar or salt is dissolved in water.

**distillation** — converting a liquid (often present as part of a complex mixture) into vapor, and then condensing the vapor and collecting the relatively pure liquid.

**distilled spirits** — ethyl alcohol separated from fermented raw materials by distillation, with such additives and dilutants as are required for a given purpose.

**distillers' grains** — the fully extracted grain materials remaining after the solubles have been removed for fermentation prior to distillation, dried and sold as an ingredient for animal feed.

**disulfide bond** — when two sulfhydryl groups (-SH) on proteins are close together, they may, under certain circumstances, react to form a disulfide bond (-S-S-) that joins two molecules or two parts of one protein molecule.

**divider** — machine that cuts masses of dough into pieces of uniform weight; there are several principles used, including filling cavities of fixed size and cutting pieces from a cylinder of dough that is being extruded at a constant rate.

## GLOSSARY OF CEREAL TECHNOLOGY TERMS 57

**divider oil** — oil used to lubricate the parts of a divider that contact dough; usually a highly refined mineral oil, although some compounds containing mostly vegetable oils are also being used.

**divinity** — or, "divinity fudge." A fudge-like confection made primarily of corn syrup and egg whites. Not baked. White, soft, and slightly aerated; usually flavored with vanilla.

**djevrek** — *Serbia/Croatia* roll of doughnut shape covered with sesame seeds.

**dobosh torte** — or, "dobos torta." A confection consisting of several thin layers of cake alternating with layers of icing, the whole coated with chocolate. Originally, the torte was coated with brown caramel, but this treatment is seldom seen nowadays.

**dockage** — the foreign material in market grain that is readily removable by ordinary cleaning devices.

**docker** — any type of utensil or machine that has the function of punching holes from the top to the bottom of dough pieces.

**docking** — punching a number of vertical small holes in a dough piece using different kinds of manual and machine implements, the purpose being to restrain puffing or expansion in the oven, thus leading to a more uniform thickness and more level surface in the baked piece. Soda crackers are good examples of the practice and effects of docking.

**doka** — *Egypt* an oiled griddle used for baking fiti, senesen, etc.

**dolce** — *Italy* sweet, dessert.

**dolci** — *Italy* pastries, cakes.

**Do-Maker** — a trade name for the continuous bread making-plant developed by Wallace and Tiernan.

**donitsi** — *Finland* doughnut.

**dora-yaki** — *Japan* bean jam pancakes, a sweet bun popular in Japan.

**dosai** — *India* also, "dosa" and "dosi." Slightly crisp, round pancake made from a fermented mixture of cereals (usually rice) and black gram.

**dosci** — *India* foods made by fermenting together rice and a pulse called black legume.

**dot** — (v) to place small bits of vegetable shortening, butter, cheese, or the like over the top of a sheet of dough.

**double-arm mixers** — a horizontal mixer including two arms, usually configured in a roughly sigmoid shape.

**double-lap traveling oven** — a large version of the traveling tray oven, in which the trays are drawn along a track that carries them both vertically and horizontally, so they can make more than two passes through the baking chamber before being unloaded.

**double-panned** — placing a cake pan containing batter on top of an inverted (empty, of course) cake pan, so as to reduce the contribution of bottom heat to the baking process.

**dough** — the term always applied to the extensible plastic mass formed from bread ingredients, but also applied sometimes to plastic cookie masses, pie crust mixtures, and the like.

**dough brake** — also, incorrectly, "dough break." Heavily built machines for pressing a sheet or large piece of dough between metal rollers that rotate fairly rapidly; its function is either to squeeze out most of the gas and orient the fibrils of the gluten as part of the developing of bread doughs and the like, or to sheet out dough and fat laminations for puff pastry and similar materials. "Brake" is the correct spelling, but "break" is seen so often that it probably should be regarded as an acceptable alternative spelling.

**dough chute** — any kind of a trough, channel, or duct through which dough descends by gravity, especially, an opening in the floor of the mixing room through which a mass of dough can be dropped to the vicinity of a depositer or divider placed on a lower floor.

**dough conditioner** — a compound, substance, or mixture that improves the physical qualities and/or processing response of a dough when a small amount of it is added to the formula.

**dough cutting wires** — in a wire-cut cookie machine, the cutting blade that is passed to and fro under an extruding die; can be either a wire or a very thin and narrow blade.

**dough divider** — see "divider."

**dough extruder** — see "extruder."

**dough hook** — for a vertical dough mixer, an agitator that is a single heavy metal rod (or cast metal part) shaped to conform to the side and the bottom of the mixer bowl. Best suited for kneading cohesive dough pieces of elastic consistency.

**doughnut glaze** — a mixture of (usually) sugar, water, stabilizer, flavor, and color applied to the surface of cooked doughnuts or similar fried products by dipping or pouring. In spite of the name, glazes are not necessarily transparent or shiny.

**doughnuts** — also, "donuts." A large and varied category of fried (in a few cases, baked) sweet dough (and batter) products, including chemically leavened, yeast-leavened, or (in a very few cases) air-leavened. The cooked products can be glazed, frosted, iced, dusted, covered with dragees or nut pieces, filled, and otherwise modified.

**doughnut screens** — screens of metal wire with lifting handles on each side; they are inserted into the deep fat fryer so that the doughnuts can be removed all at one time when they have been sufficiently cooked.

**doughnut sticks** — wooden dowels about a foot long and perhaps 0.5 inch in diameter, used for turning doughnuts in the cooker after the bottom side has been sufficiently fried.

**doughnut sugar** — a blend of powdered sugar, oil, and starch, often with some flavor such as vanillin, that is used to coat cake doughnuts.

**dough press** — a dividing machine for buns, suitable for retail bakeries, that presses a measured quantity of dough into a sheet of relatively uniform thickness and then pushes knives (dies) through the sheet to cut bun pieces of the proper weight. Also, sometimes used to refer to the quantity of dough processed in each operation.

**dough room record** — a paper form on which are entered the data generated by personnel (the mixer, scaler, etc.), as they perform their duties in and around the mixing machine.

**dough scraper** — a dull metal blade perhaps 6 inches long, with a wooden handle attached to one of the long sides, used by bakers to scrape the bench clean of adhering dough and to cut dough by hand.

**dough sheeter** — a machine consisting of two powered horizontal steel rollers, means for adjusting the space between them, and input and output conveyor belts. Used to reduce the thickness of dough sheets and to convert a mass of dough into sheeted form.

**dough splitter** — a device for cutting a slit across the top of loaves or rolls before they enter the oven; can be based on either metal knives or high pressure water jets.

**dough stretcher** — a machine that thins dough sheets or lumps by a pulling action, though roller action (analagous to rolling pin operation) is often involved in the process.

**doughy** — soft, sticky, elastic texture reminiscent of unbaked dough.

**dowdy** — a deep-dish pie or pudding, usually with a streusel-type topping and no bottom crust. The only common example is apple pan dowdy.

**dragées** — also, "dragees." (1) Sugar-coated nuts or fruits. (2) Small candy pieces pan-coated with sugar, unflavored or mildly flavored but often highly colored or coated with gold or silver foil; used for decoration.

**drawplate ovens** — ovens in which the entire hearth rolls out the front of the baking chamber for loading and unloading of loaves.

**dress** — to sift or bolt.

**dressing** — (1) Scalping off oversize particles from a flour stock. (2) A combination of firm bread pieces, spices, and broth, baked and used as an accompaniment to meats (especially roasted poultry).

**dried milk product** — any dairy product such as milk, skim milk, and cream that has been reduced to a flowable powder by spray-, roller-, or freeze-drying.

**drie-in-de-pan** — *Netherlands* a small, fluffy pancake filled with currants.

**Drinking Water Standards** — federal regulations giving detailed specifications for potable water.

**drop cookies** — cookies made by dropping portions of dough (by hand or cookie depositing machine) on to the baking pan or band.

**dropping point** — a test based on the temperature at which a sample of fat becomes sufficiently fluid to flow under the conditions of the test. A portion of molten fat is introduced into a sample cup, crystallized, and then heated at a constant rate; the dropping point is considered to be the temperature at which the sample is able to flow through an orifice in the bottom of the cup.

**drop scone** — *UK* baking powder biscuit dough cooked by dropping spoonfuls on a griddle.

**drum molder** — a machine that forms bread loaves (or other kinds of bakery products) by pressing dough pieces between a rotating drum and an outer jacket positioned to leave a channel between itself and the drum.

**dry mix** — a mixture of all (or most) of the ingredients, except water, required to make a particular bakery product. In some cases, the mix will contain only the basic ingredients such as flour, leavener, salt, sugar, etc., allowing the baker to add characterizing ingredients so as to make many kinds of products from one basic mixture.

**dry pack apples** — peeled and cored apples canned without added water.

**dry steam** — steam containing no droplets of water.

**dry weight basis** — the expression of the weight of an ingredient or product after its content of moisture has been deducted.

**dual-textured cookies** — cookies having a crisp outside layer and a chewy inner portion, the assumption being the pieces will retain this dichotomous texture until they are consumed.

**duchesse** — a European confection made of filberts, sugar, and egg whites and filled with nougat.

**duff** — a stiff flour dough with enriching agents such as suet, that is cooked plum pudding-style by boiling in a bag; plum duff contains prunes. Modern versions usually consist of stewed fruits mixed with cookie crumbs.

**dulce** — *Spain* candy; sweet.

**dull-to-dull** — term applied to the operation of roller mills when they are run so that the longer face of the corrugated cutting edge on the faster rotating roller meets the shorter face of the edge on the slower roller.

**dummy** — metal, wood, cardboard, or plastic structure shaped like a cake and used as the basis for applying decorations so the finished product can be used as a display.

**dumping** — (1) Pouring ingredients into the mixing bowl. (2) Decanting dough from the mixer into a trough. (3) Depanning baked cakes or bread.

**dumplings** — (1) Lumps of dough, not necessarily wheat flour dough; typically unleavened, designed to be cooked in a liquid (such as soup) or steamed; sometimes, strips cut from a rolled out dough, like a giant noodle, and used in the same way. (2) As in apple dumplings, a single pared apple, usually seasoned with cinnamon, etc., wrapped in pie crust dough and baked; often served topped with hot syrup.

**Dundee cake** — *UK* a cake, usually soda-leavened, containing almonds, cherries, currants, and lemon peel.

**dunst** — middlings of small particle size from which the bran has not been completely removed. The term ordinarily indicates middlings that will pass through a No. 8 or No. 9 silk, but will be retained on a No. 10 or No. 11.

**durazno** — *Mexico* peach.

**dörrobst** — *Germany* dried fruit.

**durum** — a variety of *Triticum turgidum*), formerly *T. durum*. Commercially, a type of wheat used (in the US) almost entirely as the raw material for milling into semolina that is particularly suitable for pasta; in Europe and the middle East, durum products are used in breads, particularly flatbreads.

**dust collector** — various types of equipment for removing flour and other dust from the atmosphere; one type is based on bags or other types of filters through which the air is passed, another type is the cyclone separator.

**dusting** — (1) Separating flour from middlings, especially so that the latter may be more efficiently purified. (2) Applying flour, starch, cornmeal, etc., to dough pieces and to baking surfaces to reduce sticking.

**dusting flour** — often a low grade flour that is procured specifically for dusting purposes.

**dutch machine** — an automatic cookie forming machine that forces dough into shaped cavities cut in the surface of a rotating large metal cylinder; i.e., a rotary molding machine.

**dutch process cocoa** — or, "dutched cocoa." Cocoa that has undergone an alkali treatment at some stage during its preparation — usually the cocoa beans are mixed with alkali before they are roasted. Dutched cocoa is darker and more readily and completely dispersible in water than is "natural" cocoa powder.

**dyes** — any kind of pigment that is useful, in a practical sense, for coloring other materials, such as cloth, food, etc.

## -E-

**Easter biscuit** — *UK* cookie flavored with spices and containing currants.

**eclairs** — also, "éclairs." Small oblong cakes, often made of cream puff dough, filled with either whipped cream or pastry cream and often iced with chocolate. Also, now often used for fancy iced and filled doughnuts of various shapes.

**eesti suhrukook** — *Estonia* a rich sugar cake, chemically leavened.

**e-fu** — *China* flattish, yellow, egg and wheat flour noodles, formed into loosely tangled bundles (see "skeins") and fried before sale. The consumer boils the noodles briefly.

**egg rolls** — *China* a small, thin sheet of unleavened dough is folded completely around seasoned fillings consisting mostly of vegetables (such as bean sprouts) mixed with shredded meat, small shrimp, etc.; these packets, roughly cylindrical in shape, vary from small to moderate in size and are usually fried.

**egg wash** — lightly beaten eggs or a mixture of whole eggs and milk, which is brushed or sprayed on the outside of proofed dough products just before baking so that the finished food will have a glossy brown crust.

**eggs, shell** — eggs in the shell, readily recognizable by a chicken.

**eggs, whole** — the contents of chicken eggs, with the yolk and white portions being present in the same proportions as in shell eggs. Available commercially in dehydrated, frozen, and refrigerated forms, with and without additives.

**egg white** — the white, more fluid part of the contents of a hen's egg. Available commercially in dehydrated, frozen, and refrigerated forms with and without additives. Also called "albumen."

**egg yolks** — the yellow of egg, available commercially in dehydrated, frozen, and refrigerated forms with and without additives.

**ei** — *Netherlands* egg.

**eier** — *Germany* eggs.

**eierkückas** — a pancake made of batter enriched with fresh cream, a specialty of Alsace.

**eierpannekoek** — *Netherlands* egg pancake.

**eigelb** — *Germany* egg yolk.

**einkorn** — *Triticum monococcum*, a primitive form of wheat bearing only one grain per spike; though of historical interest, it has no commercial importance in the US at the present time.

**eischnee** — *Germany* whipped egg white, meringue.

**eish shami** — *Middle East* a product resembling pita bread.

**eiweiss** — *Germany* egg whites.

**ekler** — *Serbia/Croatia* eclair, cream puff.

**elasticity** — tendency of a material to recover its original shape after it is released from a deforming force. In dough, this property is influenced by absorption, mechanical development, fermentation, and other factors. Batters normally are not elastic.

**electronic load cell** — a device used as a sensing means in scales; it depends for its action on the distortion of an electronic circuit by a load placed on the support.

**elephant ears** — a fanciful name applied to deep-fried thin discs of yeast-leavened dough, often of pizza size, coated with, e.g., sugar and cinnamon.

**elevator** — a complex consisting of large silos, fitted with devices for automatically adding and removing grain, and having facilities for transferring loads from and to, trucks and train cars.

**elote** — *Spain* fresh (or sweet) corn.

**embossing machine** — device that applies a pattern to the top of a sheet of cookie dough; variations include a rotating drum with carved impresssions on the surface and reciprocating plates bearing metal strips formed into patterns.

**emergency dough** — a yeast-raised dough prepared according to a formula that allows short-cuts in fermentation and conditioning so the batch can be finished in a relatively short time.

**emmer** — a primitive type of wheat, a variety of *Triticum turgidum* currently of no commercial importance.

**empada** — *Portugal* pie of relatively small size.

**empadao** — *Portugal* pie of relatively large size.

**empanada** — *Mexico* a kind of pie having a non-flaky dough folded over a filling such as pumpkin; generally small in size.

**empanada de horno** — *Spain* dough jacket filled with minced meat, similar to ravioli, but baked.

**empanadilla** — *Spain* small version of the empanada de horno.

**emulsifier** — (1) An ingredient that enhances the formation of relatively stable systems consisting of finely dispersed globules of fat in aqueous solutions, or the reverse. Sometimes used as a synonym for surfactant, which usage, however, lacks precision. (2) A machine that forms emulsions from fatty substances and water, either by very high speed mixing action or by impinging the ingredient mixture on a plate (homogenizer).

**emulsions** — systems of small droplets distributed throughout a continuous phase of an essential immiscible liquid. Very common in foodstuffs; mayonnaise is a typical example. It is not strictly correct to apply the term to structures consisting of gas bubbles in a continuous liquid or solid phase, although this is often done.

**enchilada** — *Mexico* in its usual form, a tortilla folded around a meat filling leaving the ends open, then baked after topping with a sauce consisting of tomatoes, chili peppers, and often cheese.

**endosperm** — the starchy white interior material of grain; this is the material the miller desires to separate and purify to yield the finest flour.

**endosperm separation index** — in milling technology, the results of a test for determining the efficiency with which a mill is removing the bran from the endosperm.

**engineered fats** — synthetic or rearranged fat molecules that are intended to have properties different from natural fats, often contributing fewer usable calories by virtue of their reduced digestibility.

**English muffin** — a yeast-raised, lean-dough bread product, in circular form, about 3 or 4 inches in diameter, with a flat top. The dough is very soft and is baked both top and bottom on a griddle or band, the spread usually being constrained by metal circles.

**enocianina** — grape skin extract, used as a "natural" coloring material for foods and beverages.

**enrich** — to improve the nutritional value of a food by adding vitamins, minerals, or other nutrients during processing.

**enriched bread** — bread made with enriched flour and containing federally prescribed amounts of certain vitamins and minerals.

**enriched flour** — wheat flour that has been supplemented with certain vitamins and minerals so that it meets FDA specifications for content of these factors.

**enrichment concentrate** — the vitamin and mineral mixture that is added to bread dough to meet the requirements for enriched bread; usually sold in the form of a powder or tablets.

**enrober** — a machine that coats the surface of a product, such as a cookie, with melted chocolate or some other fluid material that sets up when cooled or dried.

**entire wheat bread** — whole wheat bread.

**envueltos** — another of the myriad Mexican foods made by wrapping tortillas (flour or corn) around a filling of cooked, chopped (or shredded), and flavored meat.

**enzyme** — a class of proteins found in all living things that speeds up chemical reactions, i.e., an organic catalyst. There are thousands of different kinds. e.g., amylases that act to break down starch and proteases that facilitate the hydrolysis of proteins.

**epi** — also, "les epis." *France* long loaf that has been cut from the edge toward the center several times along both sides, allowing pointed chunks to stick out, a pattern said to resemble heads of grain on a stalk.

**épice** — *France* spice.

**erdnuss** — *Germany* peanut.

**ergot** — a fungus infecting certain cereal plants, particularly rye, reducing yield and posing a serious health threat to man. Historically, it has caused famine and plague, but currently of small importance in the US.

**eriste** — *Turkey* noodles.
**erucic acid** — a fatty acid of moderate toxicity found in rapeseed oil.
**escarchar** — (v) *Spain* to frost, as in escarchar el pastel, "to frost a cake"
**ESI** — endosperm separation index.
**especiaria** — *Portugal* spice.
**essence** — a flavoring ingredient, especially the highly volatile top notes of natural products, such as the aromatic substances collected during water removal from fruit juices by distillation.
**essential amino acids** — those amino acids that an animal must obtain from foods, i.e., the animal cannot manufacture enough of the compound to maintain its health and growth. Different species vary in their requirements; unless otherwise specified, it is understood that a nutritional discussion of essential amino acids refers to human needs.
**essential oil** — the type of flavoring material found in many spices and extractable by ether and similar solvents or removed by distillation; these oils are not triglycerides.
**ester** — an alcohol to which an organic acid residue has been chemically bound. Natural and artifical flavors often contain esters. The most commonly found kind of ester in foods is triglycerides, in which glycerol (the alcohol type substance) has three fatty acids attached to it.
**ethanol** — the alcohol formed by yeast during fermentation; it is also called grain alcohol; the same compound is the intoxicating component of beer, wine, whiskey, etc.
**ethoxylated monoglycerides** — a type of emulsifier made by reacting ethylene oxide with the free hydroxyl groups on monoglyceride molecules. It improves loaf volume but is not very effective in softening the crumb or increasing shelf-life.
**ethyl vanillin** — a synthetic flavor often used in imitation vanilla flavors; it is very aromatic but has a somewhat different character than vanillin.
**eutectic** — a mixture of two or more substances in a ratio such that the mixture's melting point is the lowest possible for a combination of those substances.
**evaporated apples** — peeled, cored, and sliced apples that have been dried to a moisture content of about 24%; usually treated with sulfur dioxide to prevent browning.
**evaporated milk** — unsweetened, heat concentrated milk in cans. Not much used as a bakery ingredient at this time, but has some utility for candies. Has been largely replaced by dried milk.
**evaporator** — (1) That part of a refrigeration unit in which the refrigerant fluid is allowed to vaporize and absorb heat. (2) Any equipment designed for removing water from liquids by an evaporation process.
**Eve's pudding** — *UK* sweetened apple splices topped with a spongecake batter and baked.

**expedited dough processes** — those baking schemes in which the rest or fermentation times for the dough have been reduced; for example, bread plants using brews or liquid sponges are considered to be using expedited dough processes.

**expression** — a compression process used to produce "cold-pressed" oils from natural materials.

**extensibility** — the extent to which a material may be deformed without rupture by pulling or by some other process involving the application of tension. A high degree of extensibility is the result of low yield value and high mobility associated with adequate cohesion.

**exterior-combustion ovens** — in these systems, the fuel is burned in a chamber, tubes, etc., outside the baking enclosure and the heat transferred into the oven cavity by air currents, steam tubes, or the like.

**extract** — a flavoring ingredient, ostensibly prepared by steeping a natural material, such as vanilla beans, in alcohol or other solvent, then removing all the insolubles; that which remains is the extract. Has been applied, incorrectly, to mixtures of dissolved chemicals and natural materials.

**extraction** — (1) In milling, the percentage of flour or meal obtained, using the weight of the mill mix as 100%. An extraction of 70% means that 70 lbs. of flour have been obtained for each 100 lbs. of wheat going into the mill; the percentages are usually adjusted to the same moisture content. (2) In cookie production by molding processes, the removal of the dough shape from its cavity, usually by means of adhererence to a belt. (3) In the preparation of flavor essences and the like, the contacting of a natural material with a solvent (as of vanilla beans with alcohol), then separating the fluid (extract) from the residue.

**extractive milling** — the technique of soaking hulled brown rice in certain liquids under conditions such that the bran coats are softened while the endosperm retains its initial properties, for the purpose of improving milling efficiency.

**extrude** — to force a plastic material (dough, for example) through a restricted opening.

**extruders** — a broad and diverse group of machines that are based on the principle of applying pressure to a plastic material (such as dough) in order to force it through shaped openings. Often combined with a mixing section that forms the plastic material from dry and liquid ingredients. Macaroni presses are typical examples of food extruders.

## -F-

**facultative anaerobes** — microorganisms that can grow and reproduce in either the presence or absence of oxygen.

**fadennudel** — *Germany* vermicelli, very thin spaghetti (or noodles).

**fairy cakes** — *UK* iced or frosted small rich cakes made of chemically leavened and beaten egg-containing batter. Similar to petit fours.

**Falling Number Test** — a test used to measure the level of diastatic activity of a flour; it has some value for determining the suitability of flour for breadmaking.

**family flour** — a flour designed to be used in many different kinds of products, breads, rolls, biscuits, cakes, doughnuts, etc. Not the flour of choice for a professional baker, since it does many things at a barely acceptable level, but is optimum for nothing. It is usually made from hard red winter wheat of low protein content, but in some cases from soft wheat.

**fancy patent** — a very short patent flour, representing say 60% of the total flour produced, the rest being clears, etc.

**farina** — coarse endosperm particles (middlings) from the wheat milling process. Used as the basis of a hot porridge (Cream of Wheat) and in a few bakery products. If from durum wheat, usually called "semolina." *Italy* flour.

**farina di avena** — *Italy* oatmeal.

**farina di frumento** — *Italy* wheat flour.

**farine** — *France* flour.

**farine d'avoine** — *France* oatmeal.

**farine de froment** — *France* wheat flour.

**farine supérieure** — *France* best grade of wheat flour.

**farinograph** — a small recording dough mixer that measures the changes in consistency of a dough during a prolonged mixing period. Its results give an indication of the quality of the flour or, at least, of its similarity to previous deliveries.

**fast dough** — about the same thing as an emergency dough, i.e., a dough formulated so as to ferment and condition more rapidly than a regular bread dough.

**fastlagsbulle** — *Sweden* a roll filled with almond paste and cream, a Lenten specialty.

**fat absorption** — the amount of oil a fried product takes up during the cooking process.

**fat bloom** — a dull white or grayish coating on a chocolate product; results from exposure to high (melting) temperatures, which allows certain fractions of the cocoabutter to come to the surface. Cannot be corrected except by remelting the chocolate, tempering it properly, then cooling.

**fat filters** — fine-pore sieves through which the hot fat from doughnut fryers and the like is passed periodically or continuously to remove pieces and particles originating from the fried product.

**fatias** — *Portugal* slices, as of bread.

**fatias douradas** — *Portugal* French toast.

**fatier** — *Egypt* appears to be a kind of non-sweet strudel made of many thin layers of dough coated with melted butter.

**fat rascals** — *UK* a kind of rich tea-cake made with butter or cream, and containing currants.

**fats** — In a chemical sense, triglycerides, i.e., the chemical species resulting from the esterification of one unit of glycerol with three units of fatty acids. A natural fat is a mixture of many different triglyceride molecules. Also, see "Total fat."

**fat stability** — (1) In the case of frying fats and oils, refers to the rate of breakdown of the fat, as determined by chemical analysis or by observations of foaming, discoloration, and the development of off-odors and -flavors. (2) Resistance of shortenings to the development of oxidative and hydrolytic rancidity under normal conditions of use and storage.

**fat substitutes** — synthesized or natural substances that can be used to replace ingredient fats for the purpose, usually, of decreasing the calorie content of a portion of food.

**fattiga riddare** — *Sweden* French toast.

**fatty acids** — a chemical species occurring naturally, either singly or in combination with other moieties, consisting of strongly linked carbon and hydrogen atoms in a chain-like molecule. At one end of the molecule is a reactive acid group.

**FD&C colors** — natural and synthetic materials that are on the FDA lists of certified and uncertified colors for foods, drugs, and cosmetics.

**feed** — (n) food for animals.

**feed value** — generally, the metabolizable energy per pound in a feed ingredient, when consumed by a specific kind of animal (e.g., ruminants or non-ruminants).

**feeds** — also, "mixed feeds." Prepared (formulated) rations for animals, usually consisting principally of an energy source such as corn, with added protein-rich materials and trace nutrients.

**felafel** — also, "filafil." *Middle East* Deep-fried small balls or fritters, made from bulgur, chickpeas, broadbeans, onions, lemon juice, and spices.

**fennel** — the dried ripe fruit of the perennial European herb *Foeniculum vulgare*. This spice has a mild anise flavor with a subdued menthol note, and is used much like anise, but it is not as potent as anise. The leaves and root of the plant can also be used as flavorants.

**fenugreek** — this spice is the dried ripe seed of a herb (*Trigonella foenumgraecum*) grown primarily in India and Morocco. It is used as the basis of most of the imitation maple flavors that find considerable application in icings for pastries, cakes, and doughnuts.

**fer à cheval** — *France* horseshoe-shaped loaf.

**ferment** — (1) (v) To undergo the microbiologically mediated reactions that result in the production of carbon dioxide and ethanol as well as many other physical and chemical changes in yeast-leavened doughs. (2) (n) A pre-mix similar to a broth; a fluid mixture of flour, water, yeast, and other ingredients that can be handled by liquid transfer techniques and fermented under controlled conditions until it is suitable to be used as part of a bread dough.

**fermentable solids** — those materials that can be metabolized by baker's yeast to form carbon dioxide and ethanol, the most common being sucrose, glucose, and fructose.

**fermentation** — in baking, the complex series of microbiological changes that generate carbon dioxide and ethanol in a dough or broth. Also, steps in a dough preparation process that are introduced to encourage or accelerate such changes.

**fermentation loss** — the reduction in weight occurring in a dough as the result of the change of sugars into carbon dioxide that escapes from the dough; some of the ethanol is also lost to the atmosphere, and possibly other substances.

**fermentation room** — a room or other space in which temperature, and usually humidity, can be controlled so that sponges or doughs can be uniformly fermented.

**fermentation tolerance** — the length of time a dough can be processed before or after its fermentation optimum and still yield good quality bread.

**fermented red rice** — *China* a flavoring made by adding a red food color to the lees remaining from the rice-wine brewing process.

**ferrous sulfate** — an iron compound that can be used for nutritionally enriching bakery foods.

**fersken** — *Denmark* peach.

**fettucine** — *Italy* pasta in strip or ribbon form, varying in width up to perhaps one-half inch, and relatively thin.

**feuilletage** — *France* (1) Puff pastry. (2) The sheeting out, folding, and re-rolling process that is repeated to make puff pastry dough.

**feuilleté** — *France* puff pastry case.

**feuillatines** — *France* strips of flaky pastry sprinkled with granulated sugar, a kind of cookie.

**fiber** — an inexactly defined group of food constituents having the unifying characteristic that they are not digested (or are poorly digested) in the human intestines, so that they contribute to the bulk of feces and encourage intestinal motility. Some so-called dietary fiber materials do not have the physical form of a fiber at any point in their existence.

**fiber supplements** — materials that can be added to foodstuffs to increase their content of dietary fiber; they are usually insoluble fractions of fruit and vegetable products, such as sugar beets and citrus fruits, but in some cases are derived from wood, cotton, etc.

**ficelle** — *France* a very thin loaf of bread about 10 inches long and weighing about 5 oz; a giant bread stick.

**ficin** — a proteolytic preparation from figs.

**fico** — *Italy* fig.

**fideo** — *Mexico* a very thin spaghetti, similar to vermicelli or angel hair.

**fidhes** — *Greece* thin noodles.

**field corn** — maize kernels of the common type used for animal feed, corn starch manufacture, and cornmeal milling; not sweet corn and not popcorn. "Dent" is the most widely grown variety of field corn.

**fig** — fruit of the fig tree, *Ficus carica*. There are many varieties, differing in size, color, flavor, texture, etc. The American baker uses dried figs as a basis for fig bar filling, and for very little else.

**fig bar extruders** — the machines that form fig bars by simulaneously extruding a central thick strip of fig paste and a surrounding case of unleavened (or with a small amount of chemical leavener) cookie dough.

**figo** — *Portugal* fig.

**fig paste** — ground dry figs, the principal ingredient in fig bar filling.

**figure piping** — forming designs with thin strips of extruded royal icing, imitation jelly, etc. Used for decorating cakes and the like.

**filberts** — an approximately spherical nut generally larger than a peanut and smaller than a Brazil nut. A very common ingredient in European pastries but very uncommon in US baked goods. Hazelnuts.

**filhó** — *Portugal* fritter.

**filler** — (1) A mill fraction added to a blend to provide bulk or weight rather than strength. (2) Anything added to a mixture for the purpose of increasing volume or weight without significantly affecting the functionality of the mixture.

**filling injectors** — devices consisting of a metering pump, a pointed hollow rod, and a hopper, together with a triggering device, used to insert jellies, puddings, etc. into doughnuts and other pastries.

**fillings** — imitation whipped cream, puddings, jams, and the like that are injected into bakery products, spread between layers, or otherwise put inside semi-finished (baked or fried) dessert items.

**film** — in packaging, any kind of very thin, flexible sheet made of plastic.

**film gauge** — the thickness of a packaging film, such as polyethylene or aluminum foil, usually given as mils (thousandths of an inch). Also, an instrument used to measure this parameter.

**filo** — *Greece* (various spellings, such as fila, phyllo, fillo) a paper thin sheet of pastry, very similar in properties and usage to strudel leaves.

**filters** — porous sheets or plates of cloth, paper, plastic, or metal used to separate solid particles from liquids or gases; also, more elaborate equipment containing such sheets.

**final proof** — the final proofing stage before the dough goes into the oven, after the piece is completely formed and in the pan (if it is a panned product). Some decorative work, such as splitting or applying wash may still be performed after the final proof and before entry into the oven.

**fines lames** — *France* assistant bakers who specialize in slitting dough.

**finger millet** — a food grain grown primarily in India, *Eleusine coracana*.

**finger rolls** — buns about 5 in long and 1 in wide, made of bread dough.

**fining agent** — a substance added to a liquid to assist in removing suspended matter, especially in brewing where colloidal preparations such as gelatin are added to beer at the onset of storage to speed up sedimentation.

**finishing department** — the area in a bakery where baked products arrive from the oven or cooler to be filled, iced, topped, coated, or otherwise made into finished products ready to be packaged.

**fire box** — a combustion chamber where fuel is burned to provide heat for an oven.

**fire point** — the temperature at which an oil sample, when heated under a prescribed set of conditions, will burst into flame and burn for at least five seconds. Fire points of most domestic salad oils and frying shortenings are around 650°F.

**first clear** — a flour portion collected after the patent flour mill streams have been diverted; may be divided into fancy clear and second clear.

**first proofer** — essentially the same thing as an intermediate proofer.

**first strike molasses** — in the preparation of sugar from cane, the liquid removed from the first crystallization step; useful as a food ingredient, being lighter in color, lower in ash, and milder in flavor than molasses from subsequent strikes.

**fistikia** — *Greece* nuts.

**fistikia eyinis** — *Greece* pistachios.

**fistikia arapika** — *Greece* peanuts.

**fiti** — *Egypt* bread in pancake shape made from a batter containing wheat flour, salt, and water; cooked on an oiled griddle.

**fladen** — *Germany* pancake.

**flädel** — *Germany* also, flädli. Thin strips of pancakes for adding to soup.

**flake** — in milling parlance, one of the kinds of particles found in feed bran; also a middlings particle that has flattened out during grinding.

**flake-buster** — a device used to break up middlings flakes.

**flaking grits** — in corn dry milling, a large-size particle of corn endosperm particularly suitable for making into corn flakes.

**flammeri** — *Germany* a pudding made of rice or semolina and served topped with stewed fruit or custard.

**flan** — (1) A tart baked in a bottomless circular metal form that has been placed on a sheet for filling and baking. (2) *Spain* a custard, usually of a firm consistency and molded. (3)*UK* pie.

**flan jellies** — colored and flavored clear jellies applied to the tops, and sometimes the sides, of flans, tortes, etc., to improve their appearance and prolong their shelf-life. Especially useful for flans with pieces of fresh fruit on the top.

**flapjack** — pancakes, especially thick, coarse pancakes. *UK* (1) Pancake made principally from oatmeal. (2) A fruit turnover.

**flash heat** — when the oven is empty of product but being heated, all of the materials of construction in and around the baking chamber tend to reach a higher temperature than exists when the oven is running with a full load of product; this is "flash heat" which will not normally be indicated by the thermometers (they measure only a restricted area in the baking zone) but it may cause an excessively rapid transfer of energy to the product during the first several minutes after pans start going through the oven, before thermal input and output have stabilized. It can cause overbaking of the first loads if not taken into account when specifying baking conditions.

**flash pans** — containers of water placed in the oven to absorb flash heat.

**flash point** — the temperature at which an oil sample, when heated under a prescribed set of conditions, will emit a brief flash when a flame is passed over its surface. Flash points of typical domestic oils are around 610°F.

**flatbread** — any type of food formulated, processed, and used like bread, that is baked in a relatively thin layer, with much crust and little crumb. There are innumerable examples made throughout the world. It is probably best to restrict this usage to foods made from yeast-leavened or sourdough leavened products.

**flat icing** — a simple icing composed of water, sugar, and a stabilizer such as gelatin, usually not cooked. Coloring and flavoring ingredients are also often added. Such icings are applied to cinnamon rolls and the like.

**flauta** — a tortilla rolled into a cylinder of small diameter, usually containing seasoned shredded meat, then further cooked as by baking or frying; usually served topped with a sauce..

**flavones** — plant pigments, 2-phenyl chromones; exhibiting hues yellowish to brown.

**flavor** — the combination of tastes and odors perceived when a material is ingested. Imitation flavors are those which give an effect similar to a natural product, as foenugreek seed is an imitation maple flavor.

**flavor extract** — natural and/or synthetic materials dissolved in alcohol and intended to duplicate a natural flavor, such as raspberry.

**flavor precursors** — those compounds in foodstuffs that have very little flavor or odor in the raw state, but which react during cooking to give flavored and odorous substances.

**flekice** — *Serbia/Croatia* small flat noodles.

**flensje** — *Netherlands* thin pancakes, relatively small in size.

**fleurage** — *France* dusting material (such as cornmeal) applied to the bottom of dough pieces to keep them from sticking to implements such as the peel.

**fleurons** — *France* small shapes such as fans, flowers, or geometric designs cut from puff pastry dough and baked, also sugar wafers in similar forms; fleurons are used for decorating various dishes (e.g, servings of ice cream).

**flint corn** — a type of field corn, in many respects like dent corn, but with hard, horny, rounded, or short and flat kernels, having a soft and starchy central endosperm entirely enclosed by a hard outer layer.

**flode** — *Denmark* cream.

**flodekage** — *Denmark* pastry topped with whipped cream.

**flodeost** — *Denmark* cream cheese.

**flodeskum** — *Denmark* whipped cream.

**floodability** — the property of a powder or granular material that causes it to move in erratic bursts when being delivered from a hopper.

**floor time** — the period between removal of dough from the mixer and its processing by the divider.

**florentines** — small, thin disc-shaped confections consisting usually of chocolate or toffee, sometimes covered with nuts or pieces of candied fruits. Another form is a cookie containing nuts and dried fruits, having the top covered with chocolate decorated with wavy lines.

**flour** — when not modified, the word means a finely ground material composed mostly of the endosperm of the wheat berry. There are many other kinds of flour from different varieties of grain, and the word has also been used by analogy to describe finely powdered materials not derived from grain, e.g., "flour salt."

**flour corn** — also, "soft corn." A variety of field corn, having kernels shaped like those of flint corn but composed largely of soft starch.

**flour dusters** — devices that coat pieces of dough with flour or starch; there are mechanical versions (usually based on vibratory and brushing action for removing the flour from a hopper) and pneumatic types.

**flour ferment** — a preferment or liquid sponge in which about 70% of the flour and all of the water, as well as other ingredients, are fermented together before they are added to the remaining ingredients.

**flour mill** — any type of machine or factory that reduces wheat to flour and by-products; also applied to systems for grinding other grains.

**flour salt** — pulverized common salt (sodium chloride).

**flour streams** — the many different flour-grade fractions coming off the grinding and sieving processes in a flour mill. Each stream is separately conveyed, generally in a pneumatic tube, but they can be combined in different ways to get different kinds of flour: patent, clear, straight, etc.

**flour weight basis** — a method of expressing the amount of ingredients in a formula by taking the amount of flour as 100%. Thus, a formula describing a batch containing 60 lbs of flour, 30 lbs of sugar, and 15 lbs of fat would be expressed as 100% flour, 50% sugar, and 25% fat. This old-style method of formulating should generally be avoided.

**flow sheet** — chart indicating the machines and/or the unit operations involved in a manufacturing process, arranged in a manner illustrating their proper sequence in the line.

**flue** — chimney; duct or conduit for transferring products of combustion.

**fluff** — beaten, aerated egg white and sugar to which crushed fruit has been added. Also, "fruit fluff."

**fluidity** — the reciprocal of viscosity.

**flummery** — *Ireland* There are many variations, but the traditional Irish flummery is most often a cold dessert based on cooked oatmeal, frequently in the form of a soft, jelly-like fluid. Not leavened. The earliest type of flummery appears to have been an acidic jelly made from the husks of oats.

**flûte** — *France* a loaf about 16 oz in weight and 15 to 20 in long.

**flying sponge** — a sponge mixture (used in a sponge-and-dough bread process) that is fermented for a relatively short period of time, usually with the fermentation reactions accelerated by higher temperatures, more yeast, or other methods.

**foam** — any soft mass of finely divided bubbles that has some stability, but in baking particularly applied to an aerated mass of egg and sugar, as in sponge cake batter before the flour is added.

**foam cakes** — cakes based on a foam prepared by whipping egg whites and sugar; into the foam, flour and the other ingredients are folded. Angel food cake is an example.

**foam icings** — egg white foam or some other highly aerated preparation mixed with a heated blend of syrups and other ingredients to form a light fluffy spread or fillig.

**focaccia** — *Italy* (1) Round flat bread product similar to a baked yeast-leavened pizza crust without toppings, often flavored with spices mixed into the dough and sprinkled with olive oil. Used as a bread. Also, ring-shaped cake made of fairly rich yeast-leavened dough.

**focaccia di formaggio** — *Italy* cheesecake

**foil** — a very thin sheet of metal, especially of aluminum in thicknesses of, for example, about 0.00035 in.

**foil-paperboard pans** — a construction of aluminum foil laminated to a stiff layer of card stock and formed into a pan to serve as a distribution container and baking utensil for brown 'n serve rolls and the like.

**fold** — (1) The operation which, with doughs, consists of lapping a sheet of dough over itself, and with batters consists of lifting and turning over a portion of the mass so as to lightly incorporate ingredients (as in making angel food cake batter). (2) The extent to which a flavor extract has beem concentrated as compared to some standard, e.g., 2-fold vanilla extract.

**folded oil** — an essential oil from which part of the terpene fraction has been removed, as by distillation. Thus, a 2-fold oil has had half of its "inert" components removed.

**folhado** — *Portugal* a kind of sweet puff pastry.

**fondant** — a base for icings and chocolate bonbon centers prepared by cooking sugar syrup in such a way that very small and uniform sugar crystals are formed in a continuous phase of saturated sugar syrup. The best examples appear as very white, creamy, smooth pastes.

**fondant icings** — icings with a smooth dense texture and and candy-like eating qualities; widely used for icing cakes.

**fondant sugar** — sucrose with a particle size less than about 44 microns in the largest dimension, that has been mixed with a stabilizer such as maltodextrin; used as the major ingredient in certain icings, confectionery, etc.

**Food Chemical Codex** — a group of internationally accepted specifications for food additives.

**fool** a type of dessert usually consisting of a puree of cooked fruit folded into swirls of whipped cream, sometimes with fragments of cake included.

**forårsrulle** — *Denmark* egg roll of the Chinese type.

**forced convection ovens** — baking chambers in which the hot air is circulated by fans to accelerate heat transfer to the products.

**force-deflection systems** — a mass-measuring principle used in some weighing scales based on determining the extent to which a spring or other elastic element is distorted by an applied force.

**foreign material** — any unexpected or unwanted substance found in a material described by another term, as pebbles in a load of grain, or wheat in a load of pebbles.

**forgácsfánk** — *Hungary* a fried unleavened biscuit made of a dough rich in eggs and sour cream, but not very sweet; the dough is sheeted thin, cut into rectangles, and folded in a special way before it is fried in deep fat.

**forking** — also, fork splitting. An operation used by bakers to partially separate the top and bottom halves of English muffins by pushing thin blades or tines from the sides toward the center. The consumer completes the separation by manually tearing the halves apart.

**form-fill-seal machine** — a type of packaging equipment that forms pockets from roll stock, fills flowable food products into the pocket, seals the pocket, and cuts the pouch from the roll.

**formula** — essentially, a recipe that has been adapted for large scale use in the bakery. Contains a list of the ingredients with limited descriptions and their respective amounts or percentages. May or may not include processing instructions. More generally, any kind of large-scale recipe used in industry, commerce, and laboratories.

**formula balance** — an expression indicating that the ingredients in a formula are in the correct proportion (according to an expert's predetermined standard) to yield a top quality product.

**formula feed** — a mixture of materials designed to supply a full complement of nutrients to a given type of animal (e.g., hogs, cattle, or chickens); they usually include an energy source such as corn endosperm, a protein concentrate such as soy meal, and trace nutrients.

**forno** — *Portugal* oven, "no forno"= baked. *Italy* "al forno"= baked.

**forto** — *Italy* hot in taste, spicy.

**fortune cookies** — a circle of lean dough, which may be unleavened or slightly chemically leavened, is baked until it is slightly brown but still flexible, a small piece of paper with a "fortune" printed on it is placed on the dough, and the dough folded around it into a shape reminiscent of a tricorn hat.

**fougasse** — also, "fougace." *France* Fancy shapes made from a dough of the brioche type, but somewhat lower in butter content and relatively high in sugar content.

**four** — *France* oven; *au four* means "baked."

**foxtail millet** — a type of millet grown for food in Russia, China, India, and elsewhere; *Setaria italica*

**foxy** — describes a baked product, particularly bread, that shows a tendency toward excessive redness of crust color.

**fractions** — substances separated from a mixture according to some specific operation; fractions themselves are often mixtures of compounds.

**fractionation** — in processing edible oils, the word means controlled crystallization and partitioning techniques using solvents, detergents, or cold dry-pressing to separate the semi-finished material into hard and soft fractions. Includes methods for making winterized oils, high stability frying oils, and cocoabutter replacements.

**fradzola** — *Greece* white bread, similar in composition to the U.S. counterpart.

**frangipane** — also, "frangipani." A pastry cream made with milk, sugar, flour, eggs, and butter, mixed with crushed macaroons or ground almonds; used to fill or top various cakes, pancakes, etc.

**franskbrod** — *Denmark* white bread. (*Sweden* franksbröd).

**frasage** — *France* the initial mixing of ingredients to make dough.

**frasvåffla** — *Sweden* waffle.

**free fatty acids** — fatty acids that are not chemically bound to other moieties such as glycerol. The amount of free fatty acids in an oil is measured by titrating the oil with standardized alkali in alcoholic solution.

**free-flow additives** — materials added to powders to prevent them from caking in storage and from bridging when flowing from hoppers; the 3% or so of starch normally added to powdered sugar is an example.

**freezer burn** — the loss of moisture in some parts of a frozen product due to migration of water in the vapor phase. Can occur in the absence of any melting. Usually evidenced by a change in color and texture.

**freezing** — (1) Converting water to solid form (ice) by lowering its temperature; by analogy the term is often applied to other low temperature solidification processes. (2) A form of food preservation based on lowering the object's temperature to about 0°F; this does not necessarily convert all of the liquid components to solids.

**French bread** — an unsweetened crusty yeast-leavened (or sourdough) hearth bread made from a lean dough.

**French knife** — a long knife with a pointed blade used for cutting cakes, doughs, and nuts.

**French pastries** — usually applied to an assortment of several kinds of petit fours, cream puffs, and other fancy bakery desserts, from which the customer can make a selection.

**French toast** — slices of bread coated with thin egg-milk-flour batter, then fried in shallow fat (as in a skillet); served, usually, with butter and syrup.

**friable** — easily crumbled, pulverized, or reduced to powder.

**friction factor** — in the calculation of the amount of cooling needed to maintain proper dough temperature during mixing, friction factor is a value representing the heat input resulting from the mixing energy absorbed by the dough.

**fried goods** — any dough or batter product cooked by immersing (at least partially) in hot oil.

**fried pie** — a portion-size pie typically made by folding dough over a filling, sealing the edges, and frying. Usually about 4 to 6 ounces in finished weight. Sometimes glazed.

**friterad** — *Sweden* deep-fried.

**frites** — *France* often used to mean fried potatoes.

**frito** — *Spain* fried, or to fry.

**frittaten** — *Germany* pancakes that have been shredded or sliced into narrow strips for addition to soups, etc.

**frittella** — *Italy* fritter or pancake often filled with ham and cheese or with an apple. At least in the U.S., sometimes applied to a kind of vegetable omelette.

**fritter** — originally, slices of apple (or the like) that had been coated with batter and skillet fried; now a deep-fried ball of dough, usually leavened with soda, often containing corn kernels, other types of vegetables, fruits, and/or spices. May be topped with maple syrup or the like.

**fritto** — *Italy* deep-fried.

**fritto alla milanese** — *Italy* a food piece breaded and deep-fried.

**fritura** — *Spain* fry.

**friturekogt** — also, "friturestegt." *Denmark* deep-fried.

**front of the mill particles** — relatively coarse particles removed near the beginning of the flour milling process.

**frosting** — an icing applied to the tops of cakes and pastries.

**frucht** — *Germany* fruit.

**fructosan** — a polymer of fructose,

**fructose** — levulose, fruit sugar. A reducing hexose having the same atomic composition as glucose but with a different molecular configuration. It is fermentable by bakers' yeast.

**fruekager** — *Denmark* ladyfingers.

**frugt** — *Denmark* fruit.

**frugtkage** — *Denmark* fruit cake.

**fruit cake** — a term applied to a tremendous range of products, all of them characterized by their high content of candied fruits, raisins, nuts, etc. In most of them, the crumb structure that holds the fruit together in one mass is of the poundcake type, but there are many other options. The ratio of batter to fruit/nuts is highly variable.

**fruited rolls** — bread rolls containing raisins or chopped candied fruit.

**fruit-filled bars** — fig bar analogues in which the filling contains raisins, dried apples, or blueberries, etc,

**fruit pie** — dessert products made by putting a mixture of fruit, sugar, and other ingredients into an unbaked pie crust and covering it either with another complete crust or with a latticework of dough, then baking it. Usually multiserving size.

**fruit puree** — thoroughly cooked fruit (usually including a reduction in moisture content) that has been strained or milled to give a paste or mush of uniform consistency.

**fruit syrups** — (1) Sweetener syrups to which fruit flavors have been added, to be used as table syrups, etc. (2) Juices from grapes, pears, and the like that have been condensed after removing most of the coloring and flavoring substances; they are offered as "natural, non-sugar" sweetening ingredients.

**frukostflingor** — *Sweden* ready-to-eat breakfast cereal such as corn flakes.

**frumento** — *Italy* wheat.

**fruta escarchada** — *Spain* candied fruit.

**frutta candita** — *Italy* candied/crystallized fruit.

**frying** — a cooking method that relies on hot fat as the heat transfer medium. In the bakery trade, the term usually refers to deep-fat frying.

**frying fats** — fats or oils having properties fitting them for use in frying, and particularly for deep-fat frying; some of these properties being low flavor and light color, high flash point, resistance to oxidation under conditions of use, and low viscosity.

**fu** — *Japan* washed gluten dough. Various forms are sold in stores for home use and as components for packaged foods.

**fubá** — *Portugal* cornmeal.

**fudge** — soft, creamy confections, originally made of sugar, milk, butter, and chocolate, but now of more variable composition due to the availability of stabilizers, etc. A distinguishing feature is a pleasant smooth texture in the mouth that results from the extreme fineness of the sugar crystals.

**fudge icing** — a rather dense icing usually prepared from icing sugar or fondant mixed with margarine or melted shortening.

**fuldkornsbrod** — *Denmark* whole wheat bread.

**full-complement grouper** — a machine for assembling groups of pans to fit the available space on an oven tray or rack; its distinguishing feature is that it does not deliver the pans to the oven until a full tray load is in place.

**full proof** — a proof fermentation that is allowed to continue until the piece has reached the maximum size it can sustain without collapsing.

**fumigation** — applying insecticidal smoke, vapor, or gas to an room or other enclosed space.

**fundukia** — *Greece* hazelnuts.

**fungal amylase** — an enzyme derived commercially by growing certain types of molds in a pure culture, extracting them, and isolating the desired enzyme from the extract. Used to replace the diastatic effect of malt, etc.

**fungal enzymes** — enzymes (including amylases) that have been prepared by extracting and purifying fractions of a culture medium in which molds have been grown.

**fungicides** — chemicals that kill mold cells.

**fungistats** — chemicals that halt growth and reproduction of mold cells.

**fun gor** — *China* a type of filled dumpling consisting of thin sheets of wheat-starch dough folded over chopped meat, etc.

**fungus** — a filamentous plant (often microscopic) lacking chlorophyll and reproducing by spores. Mold.

**furcelleran** — one of the gums (stabilizers) obtained from seaweed extracts.

**furfural** — a chemical intermediate and solvent, in form a colorless oily liquid; derived from vegetable material containing pentosans.

**F value** — in canning technology, defined as the number of minutes needed to kill a stated number of microorganisms at a specific temperature.

**fyld** — *Denmark* stuffing.

## -G-

**gajar** — *India* carrot.
**galactose** — a monosaccharide that in lactose (milk sugar) is chemically combined with glucose. Not commercially available as an ingredient sugar.
**galapong** — *Philippines* ground rice and water dough, left to ferment overnight, used is a basis for sweet snacks such as bibingka.
**galette** — *France* a round puff-pastry loaf, about 0.5 lb, traditionally consumed on Twelfth Night.
**gal-gal** — *India* lemon.
**galleta** — also "galletica". *Spain* cookie; sometimes cracker.
**galleta de nata** — *Spain* sandwich cookie.
**gallon** — a standard measure of volumetric capacity. The U.S. gallon is equal to four quarts of 32 fl oz each, or 3.7853 liters. The imperial gallon (UK and Canada) is equal to 4.5460 liters.
**galushki** — *Ukraine* fluffy dumplings, usually served with sour cream and onions.
**galuska** — *Hungary* bite-sized dumplings made of rough-textured cereals or white flour.
**ganache** — an icing usually formed by adding sweet chocolate shavings to boiling cream, in about equal portions; the icing is applied to cakes and the like by pouring or dipping methods while it is warm.
**gansito** — *Mexico* chocolate-covered cake stuffed with strawberry and cream flavored filling (a trade name).
**garibaldi** — a cracker/cookie prepared by inclosing a thin layer of raisins or other dried fruit between two thin dough sheets, baking the combination, and cutting the sheet or strip into rectangles of uniform size.
**garlic** — the bulbs of the plant *Allium sativum* L., which is related to the common onion. There are white, pink, and yellow varieties, but the white type is generally favored. So far as baked products are concerned, garlic bread, breadsticks, and pizza toppings are the main applications of this spice. Rather common as a constituent of snack chip dusting powders.
**gas chromatography** — in this method of analysis or separation, samples are vaporized and carried by a non-reactive gas through tubes containing solid absorbent materials. Components of the sample gases are retained by the absorbent to varying degrees depending upon the extent of their chemical affinity. This results in the components exiting the distal end of the tube at different times, at which point they can be detected by various means and their quantity estimated.
**gassing power test** — a measurement of diastatic activity in which a small amount of dough is fermented in a hermetically sealed container having a gas pressure gauge attached. The pressure rise over a period of, say, 5 or 6

hours indicates how much fermentable sugar is being produced by amylolytic enzymes.

**gastronomy** — the art of good eating.

**gaufrette** — *France* sugar wafer.

**gauge rolls** — sets of sheeting rollers, the function of which is to reduce a strip of dough to a required thickness.

**gâteau** — *France* cakes and pastries in general; also, a specialty bread or loaf, very often of a particularly rich kind.

**gebäck** — *Germany* pastry.

**gebacken** — *Germany* baked.

**gebak** — *Netherlands* cake or pastry.

**gebakken** — *Netherlands* fried.

**gebroogde pruim** — *Netherlands* prune.

**gehun** — *India* wheat.

**gel** — a material that is firm or semi-rigid, but which can be deformed with slight force; made from appropriate liquid mixtures by cooling, adding a gum or calcium, or applying some other treatment. Jell-O desserts are typical food gels.

**gelatin** — a substance obtained by extracting collagenous proteins from animal skins or bones; used for thickening or gelling many different aqueous solutions. A distinguishing feature is the transition of its solutions from thin liquids to solid but short gels over a fairly narrow temperature range (as from room to refrigerator temperatures). Unlike pectin, it does not depend on high sugar concentrations for gelling power.

**gelatinization** — when said of starch granules heated in the presence of excess water, indicates the more or less complete hydration of the starch molecules, a process that forms the typical starch gel.

**gelatinize** — to cook starch in water until the granules swell and form a viscous sol.

**gembre** — *Netherlands* ginger.

**gengibre** — *Portugal* ginger.

**Genoa cake** — a glazed cake usually based on genoese sponge and containing almonds, pistachios, filberts, and/or other nuts as well as candied citrus peel and cherries. *UK* Rich currant cake with almonds on the top.

**genoese sponge** — (*France* pâte à génoise) a very rich, air-leavened cake; contains a large amount of whole eggs and a considerable amount of melted butter.

**gepaneerd** — *Netherlands* breaded.

**germ** — that part of the seed from which the new plant starts to develop; the embryo.

**germa** — *Serbia/Croatia* yeast.

**germicide** — a chemical preparation that kills microorganisms.

**germinating compartments** — in a malting plant, bins in which moistened barley grain is allowed to remain while the acrospire and rootlet form. Usually large in size, rectangular in shape, and with provisions for occasional turning of the grain and for passing conditioned air through the contents.

**germ meal** — the contents of a mill stream, or a combination of mill streams, consisting primarily of the germs of wheat (or some other grain) that have been reduced to small particles.

**geroosterd brood** — *Netherlands* toast.

**gersl** — *Serbia/Croatia* barley.

**gerst** — *Netherlands* barley.

**gerste** — *Germany* barley.

**gezouten** — *Netherlands* salted.

**ghala** — *Greece* milk.

**ghalaktobureko** — *Greece* flaked pastry filled with custard and steeped in syrup.

**ghee** — also, "ghi." *India* Clarified butter, butterfat.

**ghlika** — *Greece* candy.

**ghlika tu tapsiu** — *Greece* any kind of a pastry cooked on a baking sheet.

**ghliko** — *Greece* sweetish.

**gibanica** — *Serbia/Croatia* thin layers of pastry alternated with crumbled cheese.

**gill** — *UK* a volume measure equivalent to, generally, 5 fl oz, or one-fourth of an Imperial pint. In the US, 4 fl ounces. Other definitions are given in the older literature. Technical reports seldom refer to this measurement.

**ginger** — the root (actually, in a botanical sense, the rhizome) of a plant, used either in candied form as a garnish or confection, or in the dried and ground form as a spice in many bakery products, but as a predominant flavor in only a few, such as gingerbread and ginger snaps. Has "heat" like pepper, but also has a pleasant aroma and taste.

**ginger biscuit** — *UK* ginger snap.

**gingerbread** — includes a wide variety of products, some yeast-leavened and some chemically leavened. The former type is seldom, if ever, seen in the U.S., but is common in Germany and some other European countries. The unifying factor in all these is the use of ginger to provide the predominant flavor. Gingercake is a less dense version of the type. American gingerbread is almost always soft, of rather high specific volume, and contains fairly large amounts of molasses.

**gingersnap** — a thin, circular, crisp cookie, brown in color (due to inclusion of molasses) and highly flavored with ginger.

**gingersnap crusts** — cream-pie crusts, formed by pressing gingersnap crumbs, shortening, and other ingredients in a crust-shaped mold.

**gipfel** — *Germany* crescent-shaped roll.

## GLOSSARY OF CEREAL TECHNOLOGY TERMS

**gittertorte** — *Germany* almond cake or tort with raspberry topping.
**glace au four** — *Sweden* similar to baked Alaska.
**glace fruit** — fruits or pieces of fruits treated with concentrated sugar solutions so that they achieve considerable storage stability while at the same time retaining some of their original flavor and appearance.
**glarus** — *Swiss* flat, round bread.
**glasering** — or "glasyr." *Sweden* Icing or frosting.
**glassine** — a semi-transparent paper (not waxed) of moderate strength and stiffness.
**glasur** — *Denmark* sugar icing or glaze.
**glaze** — a transparent or translucent coating that can be applied to bakery products either as a type of icing or as a coating for the purpose of keeping exposed fruit pieces from drying out.
**glazirovka** — *Russia* icing.
**gliadin** — a type of protein that constitutes part of the gluten fraction of wheat flour.
**glucan** — a polymer of glucose.
**glucoamylase enzyme** — basically a commercial term applied to a purified enzyme preparation used as a final converting agent in the manufacture of corn syrup.
**glucono-delta-lactone** — a form of gluconic acid in which the carboxyl group is chemically combined and thus relatively inert. When it contacts water, under certain conditions the carboxyl groups are slowly exposed and become reactive. Has been used as a slow-acting leavening acid.
**glucose** — (1) The hexose D-glucose. (2) Corn syrup — this terminology should be avoided to prevent confusion since corn syrup contains many constituents besides the hexose in question.
**glucosidic cleavage** — the splitting apart of a glucose polymer, such as starch, by the chemical addition of water molecules to the glucosidic bond holding adjacent glucose units together. Facilitated by acid conditions and by enzymes.
**glucuronic acid** — a monoacid derived from glucose, $C_6H_{10}O_7$.
**glutathione** — a reducing substance present in considerable amounts in yeast cells, and also in other natural substances. Because it can convert the disulfide bonds between protein molecules to sulfhydryl bonds (which do not join the molecules together) it can weaken gluten. This effect is useful in softening excessively strong doughs and reducing mixing times.
**gluten** — (1) The mixture of proteins in wheat flour that form, upon the addition of water, the elastic structure responsible for the peculiar structure of bread doughs. There are also other proteins in wheat flour. (2) A commercial preparation made by washing the starch and most other nonprotein components out of a wheat flour dough, leaving a rubbery, grayish mass that can be used as a intermediate in several food products (see "fu")

or dried and ground to give a powder useful as a strength-inducing additive to bread doughs, or as a nutrient. (3) See "corn gluten."

**gluten-free bread** — a simulation of yeast-leavened wheat-flour loaves in which the wheat flour has been replaced by other starchy materials and structure-building proteins not obtained from wheat, rye, barley, etc.

**glutenin** — one of the two kinds of proteins that form gluten and give it strength.

**gluten quality** — the suitability of the protein component of a particular sample of wheat or wheat flour for the production of baked products, particularly yeast-leavened white bread. A rather inexact term.

**gluten separation** — generally refers to processes such as gluten washing.

**gluten washing test** — a traditional and nearly obsolete test for flour strength; it consists of first forming a flour and water dough, then kneading the dough continuously in a stream of water so as to remove nearly all the non-gluten constituents. The remainder, containing essentially all the gluten from the original flour, can be dried and weighed or otherwise evaluated.

**glyceride** — an ester of glycerol.

**glycerine** — commercial preparations of glycerol, which see.

**glycerol** — chemically, a three-carbon polyalcohol; a sweet, viscous, hygroscopic liquid obtained commercially as a by-product of soap manufacture; the human intestine breaks down fats into glycerol and fatty acids.

**glycerol monostearate** — an emulsifier and crumb softener.

**glycolipid** — an organic compound in which a carbohydrate moiety is chemically combined with lipid.

**glycoprotein** — an organic compound in which a carbohydrate moiety is combined with a protein.

**glycyrrhizin** — a colorless crystalline glucoside, the sweet constituent of licorice root.

**gnocchi** — *Italy* little knobs (dumplings) of flour, semolina, or polenta, often fried or otherwise secondarily cooked, and served in many ways. Gnocchi can also be made from mashed potatoes.

**gohan** — *Japan* (1) boiled rice; (2) a meal.

**Golden Syrup** — *UK* a proprietary sugar syrup well known in England; described by its manufacturer as a super-saturated solution of refined sugars together with non-sugars that give it color, flavor, and other properties characteristically its own.

**goma** — also, "gomashio." *Japan* sesame seed.

**goma-abura** — *Japan* sesame seed oil.

**gomme** — *Turkey* a flatbread made from a stiff dough containing flour and milk. The unleavened dough pieces (about 15 inches in diameter and 1 to 2 inches thick) are baked between a lower hot stone and a thin metal plate with hot ashes on its top.

**Goodway mixer-foamer** — one of several similar devices that supply intensive agitation to batters and other mixtures passing through a relatively small circular chamber within which are rotating one or more toothed discs.

**gorda** — *Mexico* a masa patty, much thicker than a tortilla and perhaps 3 to 4 inches in diameter, cooked on a griddle then split and filled with meat and sauce or other savory ingredients.

**gordita** — *Mexico* small gorda.

**gougère** — *France* small puffs made from a pâte à chou mixture containing a large percentage of Gruyere cheese, joined together in a ring and baked.

**goûters fourrés** — *France* a kind of cookie made from hard or semi-hard doughs, topped with sugar or salt and filled.

**gow gees** — *China* fried pork dumplings.

**gözleme** — *Turkish* small pieces of non-sweet puff pastry made of flour, butter, and egg yolks, fried, dipped into sweet syrup, and sprinkled with nuts.

**grädde** — *Sweden* cream.

**gräddfil** — *Sweden* sour cream.

**gräddtårta** — *Sweden* sponge layer cake with cream and jam filling.

**grading** — (1) The practice of applying a numerical or descriptive score to grain or the like, according to an established set of specifications. (2) The tendency of grain or other material inclosed in a bin, to separate according to density and particle size. Lighter particles tend to settle near the bin wall during filling of the bin, while the heavy particles tend to exit the bin first. Can cause severe problems in dispensing if not corrected.

**gradual reduction** — the modern process of milling, in which the goal is to produce middlings rather than to avoid doing so, as was the principle before the New Process was introduced. In gradual reduction, the middlings are sized and separated from the bran by sieves and purifiers, and the particles in each size range are re-processed under conditions peculiarly suited to that size. The process is repeated until the desired end products are obtained.

**Graff malting plant** — one design of rotary drum malting system, in which one drum accommodates all three stages of the malting process.

**graham bread** — whole wheat bread based on a recipe of Dr. Graham, who claimed great health benefits accrued from consuming this food.

**graham cracker crusts** — cream pie crusts formed by pressing into a pan-shaped mold a mixture of graham cracker crumbs, shortening, powdered sugar, and other ingredients.

**graham crackers** — present day graham crackers are usually made from chemically-leavened doughs containing a fairly high proportion of whole wheat flour and they are usually fairly sweet due to their content of molasses, etc. They are often formed in about the same shape and size as a soda cracker and can be made on the same kind of equipment.

**graham flour** — whole wheat flour.

**grahamleipä** — *Finland* graham (whole wheat) bread.

**grain** — when used in connection with bread quality, means the size, shape, and arrangement of the cells or bubbles comprising the crumb. For most bakery products, fine and uniform cell structure with thin cell walls is the desired grain.

**graining** — adjusting the cooling rate and stirring conditions of a supersaturated sugar solution so that crystals of the desired size will form, as in making fondant.

**gramola** — a heavy-duty rotary kneader for pasta dough.

**granero** — *Spain* granary.

**granola** — a breakfast cereal of indeterminate composition, but in most cases containing a variety of processed or semi-processed cereal chunks and flakes combined with dried fruits and nuts. There are ready-to-eat and to-be-cooked forms.

**granulated** — formed into small, fairly uniform pieces by cutting, milling, or crystallizing under controlled conditions; not a powder; granulated sugar is an example. Nuts are often "granulated" by a cutting process.

**grape juice** — ordinary grape juice is rarely used as an ingredient in any food except jellies and some confections, but decolorized, deodorized, and de-acidified grape juice is being sold as an "all natural, fruit sweetener" for general replacement of corn syrup. Purple grape juice has also been recommended as a food color.

**graphite** — one of the forms of elemental carbon. Can be used as a lubricant under very high temperatures where greases and oils are unstable, as on oven band supports.

**grasso** — *Italy* oily, rich in fat.

**gratin** — or, "gratinée." describes a casserole type of preparation covered with seasoned bread crumbs and, perhaps, grated cheese, then browned by broiling or baking.

**graubrot** — *Germany* brown bread, black bread.

**graupensuppe** — *Germany* barley soup.

**gravity feed depositor** — a depositor that depends on the force of gravity to draw batter into the cutting mechanism (primarily used for doughnuts).

**grease** — a petroleum lubricant of plastic consistency; also, a very viscous edible fat or oil.

**greaseproof papers** — paper made from cellulose fibers that have been processed so as to retain a high-moisture content after they pass through the paper-making machine. This causes them to resist the penetration of oils and fats.

**greasing** — (1) Spreading or spraying food oil or fat on dough pieces, pans, etc. (2) Applying petroleum grease to the moving parts of machinery to reduce friction and wear.

**greasing machines** — machines designed to spray or wipe fats or oils onto the baking surfaces of bread and roll pans, as a means of facilitating removal of the baked product.

**green** — a property of certain freshly milled flours that causes them to yield undeveloped doughs unless the flour is allowed to age, or the doughs made from it are given more oxidation, longer mixing, or longer fermentation than usual. Such flours tend to yield bread lacking in oven spring and having a flat top, foxy color, shiny pan crust, and coarse round cells in the crumb.

**green peppers** — bell peppers.

**griddle** — (n) a metal or ceramic plate or pan, often circular with low (or no) sides, typically used on or over a heat source for cooking pancakes and the like; also applied to the baking surfaces of large automatic ovens in which English muffins are cooked.

**griddle cups** — the rings placed on griddles to restrain the spread of English muffins during the baking process.

**griesmeel** — *Netherlands* semolina.

**griess** — *Germany* semolina.

**grillage** — *France* a flavoring and texturizing material for confections made by mixing roasted nuts and caramelized sugar, then breaking down or grinding the mass after it has cooled.

**grissini** — *Italy* breadsticks, typically about ten inches long.

**grissini sottile** — *Italy* thin dry breadsticks.

**grissini conditi** — *Italy* flavored breadsticks, a common snack item in Italy; may be flavored with fennel seeds, an oregano-pepper-basil mixture, or cheese.

**grits** — also, "hominy grits," Coarsely ground corn endosperm, used as a processing intermediate or as the principal ingredient in breakfast cereal, mush, etc.; also applied, in milling parlance, to larger particle fractions of other grains.

**griz** — *Serbia/Croatia* farina.

**groat** — what remains after the hull is removed from a grain of oats.

**grosella** — *Spain* currant.

**gröt** — *Sweden* porridge, gruel, cooked breakfast cereal.

**grouper** — a conveyor system that accumulates pans or straps to form a group of the optimal size for some operation such as filling a proofer shelf or oven tray.

**grozdjice** — *Serbia/Croatia* raisins.

**grunt** — a fruit cobbler, in which the topping is a soda-leavened biscuit dough that is cooked mostly by the water vapor developed when the underlying fruit is simmered in a closed vessel.

**grzbek** — *Poland* so-called "mushroom" (from appearance, not ingredients) cake, similar to kaiserschmarren.

**guar** — a water-absorbing and -binding vegetable gum obtained from the seed of the guar plant. Useful for thickening and stabilizing fillings, toppings, batters, etc., but not effective as a gelling agent.

**guava** — a tropical fruit used primarily for making jellies or purees.

**guchul pan** — *Korea* very small pancakes served as an appetizer with a tart, salty sauce and crushed sesame seeds, and accompanied by an assortment of eight or mre savory fillings .

**gugelhopf** — *Germany* also, "gugelhupf" and "kugelhopf." A molded cake made from yeast dough, often relatively high and narrow with a hole in the center, usually containing almonds and raisins.

**guillotine** — a straight metal strip positioned crosswise of a conveyor belt and provided with a means for imparting an up and down movement, used primarily for cutting dough sheets and coils into short pieces.

**gujiya** — *India* sweet stuffed puris.

**gula jawab** — *Indonesia* sugar.

**gum arabic** — a powder obtained from certain types of acacia shrubs and used for increasing the viscosity of liquids or (in high concentrations or in combination with other gelling agents) for forming sticky, "long" gels.

**gumming** — formation and accumulation on heating surfaces of a fat-insoluble sticky material resulting from high temperature breakdown of fats and oils. The gummy material results from oxidation and polymerization of the glycerides.

**gum arabic** — a plant exudate gum, cold water soluble, of relative low viscosity sometimes used in icings, etc.

**gum ghatti** — a plant exudate gum, a good emulsifier and film-former.

**gum guar** — a plant exudate gum of relatively high viscosity, reacts with milk, and dissolves without heating.

**gum karaya** — plant exudate gum that hydrates relatively rapidly in cold water.

**gums** — substances that, when mixed with water, give gelatinous, soft, rubbery masses, or, more generally, absorb water and thicken solutions. There are both natural and synthetic gums.

**gum tragacanth** — a natural gum suitable for increasing viscosity but not effective for gelling purposes.

**gur** — *India* sugar.

**guyaba** — *Spain* guava.

**gwaytio** — also, "kway tio." *Thailand.* rice ribbon noodles.

**gypsum** — a natural (mined) form of calcium sulfate that has been used as a mineral enrichment in bakery foods and to compensate for very soft ingredient water.

**gyüszüfánk** — *Hungary* deep-fried small thin discs of dough from flour, whole eggs, and salt. Called "thimble doughnuts," but mostly used for adding to soup.

## -H-

**hafer** — *Germany* oats.
**haferbrei** — *Germany* oatmeal porridge.
**haferflocken** — *Germany* rolled or flaked oats.
**hafergrütze** — *Germany* oatmeal.
**hafremjöl** — *Sweden* oatmeal.
**haggis** — a pudding or sausage, the major ingredient of which is oats, this grain being combined with chunks of cattle livers, hearts, lungs, etc., and with suet, salt, and other ingredients. The casing for the sausage is the stomach of the animal.
**haiga-mai** — *Japan* non-glutinous rice that has been polished in a manner such that the germ remains on the grain.
**haldi** — *India* turmeric
**half-high grinding** — see "New Process."
**half-product** — in snack food processing, an intermediate product usually in the form of unexpanded pellets of gelatinized cereal material can later be expanded by frying or other means.
**halka** — *Turkey* similar to bazlama (q.v.) except baked in a peel oven.
**halophilic** — describes microorganisms that can grow in solutions containing a relatively high concentration of salt.
**hälsokost** — *Sweden* organic health food.
**hamburger roll** — a soft round bread roll of about 4 inches diameter and 1.5 inches thickness, made from a fairly lean yeast-leavened dough.
**halva** — *Middle East* a confection made from semolina, oil, sugar, chopped almonds, cinnamon and other spices, fruit, chocolate, etc,
**hapankorppu** — *Finland* very thin rye crispbread.
**hapanleipä** — *Finland* rye bread.
**häppchen** — *Germany* similar to a canapé or bouchée.
**hard butter** — a generic term used in the food industry to describe a class of specialty fats with physical characteristics similar to those of cocoa butter. They are used in confectionery coatings and centers, etc.
**hard crack** — a stage in sugar boiling that is reached about 280°-310°F.
**har dhania** — *India* coriander
**hard meringue** — a meringue with a relatively high proportion of sugar, formed into shapes and baked only until the exterior is very slightly colored but the interior still retains some softness. Used as separate confections or as decorations on, and components of, more elaborate desserts.
**hardness** — (1) When used in reference to water, indicates the amount of certain minerals, especially calcium salts, determined by standardized methods. (2) The resistance to crushing of individual wheat kernels and/or the pattern of kernel rupture in the milling process. It appears to be

related to the degree of bonding of a certain kind of the protein to the starch granules.

**hard red spring wheat** — a type of *Triticum aestivum* from which the strongest bread bread flours used in the U.S. are milled; normally has a high content of good baking quality protein. Harvested in the autumn.

**hard red winter wheat** — the type of *Triticum aestivum* grown in greatest quantity in the U.S., and from which most of the bread flour and all-purpose flour is milled. It is not very satisfactory as a source of flour for cakes and pastries. During a single season, HRW from different sections of the country may yield flours of significantly different baking quality. Harvested in the summer.

**hard roll** — a yeast-leavened bread roll with a crisp crust.

**hard sauce** — a simple uncooked topping made of powdered sugar beaten up with butter, often flavored with rum, brandy, or the like. The consumer applies this to hot mince pies, plum puddings and similar desserts.

**hard sweets** — plain cookies made from a rather lean formula dough, often processed on a brake, sheeted thin, and baked almost to dryness. These cookies are very crisp or even hard in texture.

**hardtack** — in its original form, a dense, hard-baked food product consisting of an unleavened dough of flour and lard that had been formed into large round pieces and baked almost to dryness. Its only positive feature was its relatively long storage life, which made it a valuable ration on sailing vessels, etc. Mostly used broken up as an additive to chowder.

**hard water** — water that contains a relatively large amount of minerals; water hardness is expressed as grains per gallon or parts per million of equivalent calcium carbonate; "Hard" water has been arbitarily defined as having 120 to 180 ppm, "Very hard" water as above 180 ppm.

**har gow** — *China* small dumplings (ravioli), comprised of a shrimp paste enclosed in a semi-circular pocket, with pleats, made from a thin layer of wheat-starch dough.

**harina de avena** — *Spain* oatmeal.

**harina de trigo** — *Spain* wheat flour.

**harps** — the holder or frame for the wires used in a wire-cut cookie machine to cut the extruding dough cylinders into pieces of the desired size.

**harusame** — *Japan* noodles made from soybean flour.

**hasselnod** — *Denmark* hazelnut.

**hasselnöt** — *Sweden* hazelnut,

**hasty pudding** — a porridge, traditionally, cornmeal (but sometimes wheat flour or oats) stirred into boiling water or milk. May be served with milk, cream, and/or a sweetener.

**havermoutpap** — *Netherlands* oatmeal porridge.

**havregrod** — *Denmark* oatmeal.

**havregryn** — *Sweden* oats

**havregrynsgröt** — *Sweden* oatmeal porridge.
**havremelsbrot** — *Denmark* oatmeal bread, yeast-leavened. Contains some wheat flour.
**hay** — the entire herbage, sometimes including the seeds, of grasses and other forage plants such as legumes, harvested and dried for later use as cattle feed.
**hazelnut** — filbert.
**headmeters** — devices that measure the loss in pressure between two points in a pipe containing a flowing liquid, thus providing a basis for calculating rate of flow.
**head of mill** — that part of the mill where the initial processing of the grain takes place.
**healthful** — Serving to promote health of body or mind; wholesome; salutary.
**healthy** — conducive to health; as, a healthy exercise, climate, or food.
**hearth** — the heated baking surface of the floor of the oven; the part of the oven on which the products (or pans) rest during baking.
**hearth bread** — originally, loaves or rolls baked on the floor of the oven, without the use of pans. Now, often applied to bread or rolls baked in or on pans that do not confine their lateral expansion.
**heat economizers** — devices used, especially on fryers, to reduce fuel consumption by conveying raw products through the steam emerging from the frying vat to preheat them.
**heat exchanger** — a piece of equipment that efficiently (and usually continuously) transfers heat from one fluid (liquid, gas) to another.
**hedelmä** — *Finland* fruit.
**hedelmämehu** — *Finland* fruit juice.
**hedonic scale** — a method for subjectively rating the hedonic quality or sensory desirability of, e.g., a foodstuff.
**hefekranz** — *Germany* a ring-shaped cake.
**helote** — *Mexico* corn, generally sweet corn.
**heterofermentative bacteria** — types of bacteria that produce both lactic acid and acetic acid; they are important flavor-producing agents in some sourdoughs.
**high amylose starch** — a starch containing over 50% amylose.
**high fructose corn syrup** — or "HFCS." corn syrup in which a significant portion of the glucose has been enzymatically transformed into fructose. The percentage of fructose varies between commercial products.
**high intensity mixers** — mixers in which a high level of energy is imparted to the contents, usually the result of rapidly rotating agitators.
**high maltose corn syrup** — corn syrup in which the maltose content has been greatly increased by enzymatic processes, useful in some confections to improve storage stability.

**high milling** — grinding with the rollers widely separated, necessitating a larger number of millstands. This arrangement will ordinarily yield more or better flour in succeeding grinds, but reduces the plant throughput.

**high-protein supplements** — generally applies to ingredients for animal feeds that are used to increase the utilizable nitrogen content of the finished ration; examples are defatted soybean meal, corn gluten, and meat meal. Also, applies to materials used for human nutritional supplementation, these being mostly soy protein isolates or casein derivatives.

**high ratio cake** — this generally means a cake containing more sugar than flour; virtually all layer cakes made nowadays are of this type. Routine trouble-free mass production of this type of cake was made possible by the development of emulsifier shortenings and special cake flours.

**high ratio cake flour** — a soft wheat flour of about 8% gluten content that has had its pH lowered to 5.2 or less by chlorine treatment.

**high ratio shortening** — a shortening containing emulsifiers such as monoglycerides or polysorbates, and suitable for making high ratio cakes.

**high speed mixer** — the modern type of horizontal dough mixers with motors and gearing made strong enough to enable the mixer arms to rotate relatively rapidly; special vertical mixers are also described as such.

**higo** — *Spain* fig.

**hiivaleipä** — *Finland* yeast bread.

**hillo** — *Finland* jam.

**hillomunkki** — *Finland* jelly doughnut.

**hirse** — *Germany* millet.

**hiyamugi** — *Japan* a form of wheat-flour noodle.

**hjortetakker** — *Denmark* a moderately sweet, yeast-leavened dough, fried; Danish doughnuts.

**HLB system** — hydrophile-lipophile balance, a method for rating the emulsifying properties of a specific compound in a specific system.

**hleb** — *Serbia/Croatia* bread.

**hleb sa kimon** — *Serbia/Croatia* rye bread with caraway seeds.

**holding roll** — the slower rotating member of a pair of rollers.

**holding tank** — any large vessel in which some fluid (e.g., preferment) is held for a time after it is mixed or between any processing steps.

**hominy** — (1) Old style terminology, now rarely used in the marketplace, corn grains with the hull removed and sometimes broken into fairly large pieces. See "grits." (2) Also, "lye hominy," a processed food; whole kernels of maize that have been cooked in a weak alkali solution, then washed to remove alkali and hull, and canned. Used as a vegetable, as the basis for casserole dishes, and fried.

**hominy feed** — the principal by-product made in dry corn milling. It contains the bran, tip caps, ground cake from the oil mill, and the tailing streams from the grit reduction system.

**homogenizer** — a device that converts a mixture of fat and aqueous solution into a relatively stable suspension of tiny fat globules. Typically used to prevent the butterfat of milk from separating.

**honey** — the sweet viscous syrup made by bees; there are many different floral varieties, such as clover honey and orange blossom honey; its sweetening power is due mainly to fructose and glucose.

**honey bread** — in the US, wheat flour bread containing a specified amount of authentic honey.

**honeybuns** — in the US, deep-fried cinnamon rolls with a plain glaze, sometimes containing a little honey.

**honeycake** — various kinds of European specialties, generally cakes or thick cookies, supposedly sweetened only with honey. They are usually dense, rather tough, and fairly dark in color.

**hong mei** — *China* fermented red rice.

**honig** — *Germany* honey.

**honing** — *Netherlands* honey.

**honning** — *Denmark* honey.

**honung** — *Sweden* honey.

**hopper** — a tank or receptacle, usually with angled sides, positioned above a processing device so as to receive ingredients or intermediates and hold them briefly until the processor is able to accept them. Also, similar containers used to temporarily hold doughs, batters, fillings, etc.

**hops** — a customary ingredient in the preparation of beer and allied beverages, being the ripened and dried pistillate cones of certain vines of the *Humulus* genus, especially *Humulus lupulus*; provides a bitter flavor and has a preservative effect.

**hordein** — a protein complex that is the major component of the endosperm matrix in barley grain.

**horizontal mixer** — in breadmaking systems, a mixer with a U-shaped bowl having flat sides and an open (or openable) top; the agitator consists of thick metal arms running from side to side; shape of the agitator blades can vary, but is usually cylindrical for developing doughs; the inside of the bowl may contain one or more large ridges to help control dough movement. The bowl is usually jacketed so that refrigerant can be circulated around it to prevent excessive heat accumulation in the dough.

**hörnchen** — *Germany* roll shaped like a crescent or horn.

**horno** — *Spain* oven; "al horno" means "baked."

**hot cross buns** — sweet, spicy, fruity, buns with a cross-shaped depression on top which is filled with a plain frosting; Easter specialties.

**hot dog bun** — a yeast-leavened bread roll with moderate amounts of enriching ingredients, sized to fit a frankfurter sausage.

**hot print icer** — applies hot icing to a baked product with a roller that contacts the item.

**hot-serve breakfast cereals** — cereal products such as corn grits, wheat semolina, and oatmeal, intended to be mixed with water, heated until the granules have been penetrated by the water and the starch at least partially gelatinized, then served to the consumer.

**hot sponge** — a method of bread baking in which the sponge is fermented at a temperature perhaps 10° to 15°F higher than used in normal practice.

**hsing jen ping** — *China* Chinese almond cookies.

**htamin lethoke** — *Burma* rice, noodles, and a large assortment of condiments and additives that are combined by each diner, according to choice, using only the fingers.

**huevos** — *Spain* eggs.

**hull-less barley** — also, "naked barley." A type of barley in which the kernel threshes free (i.e., the lemmas do not adhere to the kernel), as it does in wheat.

**humectant** — a substance that absorbs moisture from the atmosphere; many sugars have this property. Net absorption ceases when the water activity of the substance reaches that of the surrounding atmosphere.

**humidifier** — a device that adds humidity (i.e., moisture) to the atmosphere of a proofing room or other space. Can be either automatically or manually controlled.

**humidity** — usually expressed as "Relative Humidity" which is an expression of the percent of moisture in air as related to the total moisture capacity of that air at a particular temperature.

**hunaja** — *Finland* honey.

**Hungarian process** — the modern flour milling system of gradual reduction with rollers and a primitive type of purifier was first widely practiced in Hungary, and so, for a time, such milling was called by this name.

**hushpuppy** — a fairly small ball of a cornmeal dough with seasonings, usually unleavened, fried in a skillet or deep fat.

**hutzelbrot** — *Germany* rich bread containing prunes and/or other dried fruit.

**hydrate** — a compound that crystallizes with a specific number of water molecules.

**hydraulic load cell** — a weight-measuring sensor that generates hydraulic forces proportional to the mass it supports and transmits the force to indicating units by means of a liquid contained in rigid tubes.

**hydrogenated fat** — a fat that has been reacted chemically with hydrogen gas in the presence of a catalyst. The purpose is to stabilize and/or harden the original fat.

**hydrogenated starch hydrolysates** — materials prepared by the chemical reduction of corn syrups, leading to a mixture of sugar alcohols having a sweet taste and (usually) reduced fermentability, decreased tendency to crystallization, etc.

## GLOSSARY OF CEREAL TECHNOLOGY TERMS 95

**hydrogenation** — the chemical operation of adding hydrogen atoms to a compound, commercially used in the food industry to modify carbohydrates and fats.

**hydrogen ion concentration** — water and all water solutions contain hydrogen ions resulting from the spontaneous ionization of water molecules; acid solutions contain more, alkaline solutions contain fewer, hydrogen ions than pure water. Usually reported as the pH, i.e., the negative logarithm (to the base 10) of the hydrogen ion concentration.

**hydrogen sulfide** — a poisonous gas having a very foul odor, found in trace amounts in some water sources.

**hydrolyzed vegetable proteins** — (HVP) flavors produced by chemical or enzymatic hydrolysis of vegetable proteins.

**hydrolysis** — a chemical reaction involving breakdown of molecules through their interaction with water. For example, an ester may hydrolyze in the presence of water, under certain conditions, to form an acid and an alcohol. The reaction can be catalyzed by the enzymes called lipases and by strong acids or alkalies.

**hydrolytic rancidity** — a staling phenomenon observed in many fat-containing products that results from the breakdown of triglycerides into glycerol and free fatty acids and results in off-flavors sometimes described as "tallowy." The process can result from enzymic activity or from nonenzymic hydrolysis

**hydrometer** — an instrument for determining the specific gravity of a liquid; a common design is a weighted bulb surmounted by a stem or rod on which graduations have been engraved. When the instrument (which is jacketed with glass or plastic) is placed in a sugar syrup or the like, the bulb will sink in the liquid to an extent dependent on the density of the liquid so that the specific gravity (or a number that can be converted to specific gravity) is read off the stem at the surface of the liquid.

**hydroperoxidases** — enzymes that catalyze the oxidation of certain aromatic amines and phenols by, e.g., hydrogen peroxide.

**hydrophilic group** — an arrangement of atoms, as part of an ion or molecule, that tends to form bonds with water molecules.

**hydropneumatic laminators** — machines for sheeting, folding, and resheeting doughs (as in soda cracker production) that rely on hydraulic and pneumatic devices (as opposed to electrical motors or levers) for energizing the individual parts of the machine.

**hydroponics** — the technology or practice of growing plants with their roots immersed in aqueous solutions of nutrients instead of in soil.

**hydroxylated lecithin** — lecithin that has been chemically treated to improve its hydration.

**hydroxyl group** — a chemical radical consisting of one oxygen atom combined with one hydrogen atom.

**hydroxypropylcellulose** — one of a group of gums based on chemically-modified cellulose, used in foods, cosmetics, etc., for increasing the water holding power of a mixture.

**hydroxypropylmethylcellulose** — a highly modified cellulose that is used to impart increased viscosity and water-retention properties to icings and the like.

**hygrometer** — a device for measuring relative humidity.

**hygroscopic** — describes a substance that absorbs and retains moisture from air that is within normal RH ranges. Glycerol and many sugars are hygroscopic, for examples.

**hysteresis** — a lagging or retardation of an effect, when the forces acting on a system are changed, e.g., when the viscosity of a solution exhibits a different rate of change when cooled as compared to the rate of change when heated, i.e., the curve in one direction is displaced from the curve going in the other direction.

**hyytelö** — *Finland* jelly.

## -I-

**ice cream cone** — a thin layer of batter that has been formed into a crisp-textured cone either by baking in a cone-shaped mold or by curling a flat sheet baked in a waffle-like mold; used to hold a serving of ice cream or, rarely, some other kind of confection. The batters are usually simple flour, starch, water, and sugar mixtures of low viscosity.

**ice cream cornet** — *UK* ice-cream cone.

**icing** — a very broad and poorly delimited category consisting of just about all sugar-containing pastes that can be spread on, poured over, or otherwise distributed on the surface of a baked product. Glazes, frostings, and even some fillings overlap this category. Most adjuncts of this class are made from confectioners' sugar, fat (e.g., butter), and liquid (e.g., water), with the addition of stabilizers, whipping agents, colors, flavors, and preservatives, as required. Meltable coatings such as chocolate and chocolate analogues are generally not considered icings but rather enrobings, couvertures, or just "coatings."

**icing knife** — a spatula with a handle designed for spreading icing on cakes.

**icing screen** — a metal screen with fairly large openings on which doughnuts can be placed while being glazed or iced; the excess glaze flows through the screen and is collected and re-used.

**icing sugar** — powdered sugar, confectioners' sugar (which see).

**idli** — *India* small patties shaped like a thin muffin made from relatively coarse rice particles and, usually, a pulse called black legume, the latter having been previously fermented; cooked by steaming. Served with, e.g, coconut chutney or spicy sauces.

**imbibitional properties** — refers in general to the propensity of a solid material to take up a liquid, but in particular to the capacity of flour or gluten to absorb water.

**imagawa-yaki** — *Japan* a chemically-leavened muffin filled with bean jam.

**imbir** — *Russia* ginger.

**imbiss** — *Germany* a snack.

**impact mills** — machines for grinding grain and the like, depending for their action on the centrifugally-induced impact on pins or chamber walls of the relatively large particles.

**impingement oven** — oven containing many vertically-oriented tubes through which hot air is blown onto the product.

**incasciata** — *Italy* layers of dough alternated with meat sauce, hard-boiled eggs, and grated cheese.

**incorporating** — blending one or more ingredients into a partially mixed batch.

**indian pudding** — traditionally a baked pudding, the usual ingredients being cornmeal, milk (or water), and molasses.

**indica rice** — one of the generally recognized subspecies of common rice, *Oryza sativa*.

**indirect fired oven** — ovens in which the basic heat source, burning fuel, is used to heat air that is then blown into the oven to bake the product.

**infant cereals** — generally, pre-cooked then dried compositions of cereal products and other nutrients, intended to be mixed with milk or water before it is fed to babies.

**inferential meters** — a measuring device for liquids that depends upon the moving of a screw, vane, or some other inertia-dependent mechanism by the flowing liquid.

**infrared analyzers** — see "infrared reflectance instruments."

**infrared oven** — such ovens transfer most of the heat energy to the product via infrared radiation; of course, all ovens (except microwave ovens) transfer some energy by way of infrared radiation.

**infrared radiation** — invisible electromagnetic radiation of relatively long wavelength as compared to light; efficient transmitter of heat.

**infrared reflectance instruments** — devices that measure the extent of absorption of a beam of infrared radiation that occurs at or near the surface of the tested material and from these data infers some aspect of the composition of the object.

**infusion** — a natural flavoring material processed by extended solvent treatment, generally by immersing the material in hot (>60°C) alcohol for a relatively long time.

**infusion mashing** — a method of preparing wort (primarily for top fermentations) that requires the conversion and extraction of the mash be carried out in a single stage and at a single temperature (compare this to "decoction mashing").

**ingefær** — *Denmark* ginger.

**ingefära** — *Sweden* ginger.

**ingrediator** — a tank in which minor ingredients are blended with a small amount of water and then portions of the mix delivered to the main batches; also called a slurry tank.

**ingredient** — a material (solid, liquid, or gas) that is intentionally added to a mixture for the purpose of achieving a foodstuff having the desired properties; ingredients are not always components, since some ingredients are lost and some are chemically changed during processing, while inadvertent contaminants may also be components.

**ingwer** — *Germany* ginger.

**injectors** — any device for pushing a material into the interior of an object, but particularly a pump and needle device used for inserting pudding-like fillings into the interior of doughnuts, cream puffs, etc.

**injera** — *Ethiopia* fermented round flatbread made from sorghum meal; the dough is fermented for 2 or more days, boiled, and steam cooked.
**inkivääri** — *Finland* ginger.
**input devices** — any device that transforms some aspect of an event happening outside the central processing unit of a computer into a set of electrical impulses that can be "understood" by the computer. A keyboard is an input device.
**instore bakery** — a bakery operating as a separate unit within a supermarket or in any other retail outlet not principally a baked goods store.
**interesterification** — a chemical change resulting in the random rearrangement of the fatty acids in the triglycerides of a natural fat. The distribution of fatty acid species in any naturally occurring oil conforms to a general pattern characteristic of that ingredient, but the distribution can be changed to a random pattern through use of a catalyst and appropriate processing conditions. A common procedure in modern shortening manufacturing plants.
**intermediate proofer** — also, "interproofers," "first proofers," and "dry proofers." In conventional breadmaking processes, equipment that receives dough balls from the rounder and allows them to "rest" for a few minutes before they go to the molder.
**intermediate proofing** — a stage in which the rounded dough pieces are allowed to ferment so that the gluten relaxes, making the dough more suitable for processing in the molder.
**invert sugar syrup** — syrup formed by hydrolyzing part of the sucrose (cane or beet sugar) content of a solution into its constituent hexose sugars, glucose and fructose. Usually made by heating sucrose syrup with acid or by adding invertase to sucrose syrup.
**invertase** — an enzyme preparation obtained from yeast and used for converting sucrose to fructose and glucose, giving a sweeter syrup of lower viscosity and greater hygroscopicity. Also, the pure enzyme.
**iodine value** — an expression of the degree of unsaturation of a fat. It is measured by determining the amount of iodine that will react with a natural or processed fat under prescribed conditions. Iodine reacts with the chemically unsaturated bonds in the fatty acids.
**iodized salt** — sodium chloride to which a very small amount of potassium iodide has been added to provide the nutrient factor, iodine, to consumers.
**iodophor** — combination of iodine with another substance that causes the slow release of iodine when the combination is contacted with water.
**irradiation** — the treatment of an object or material with high energy electrons or gamma rays for the purpose, usually, of reducing biological contamination.
**Irish soda bread** — a soda bread typically made in loaf shapes, often as hearth bread, and traditionally with buttermilk as the acid-reacting com-

ponent of the leavening system. Seldom sweetened, but occasionally includes raisins or the like.

**irmik helvasi** — or, "irmir helvasi." *Turkey* Pudding based on semolina that has been fried before it is mixed with the liquid (water and/or milk); other ingredients are sugar, butter, and almonds.

**isolate** — (n) a flavoring or aromatic material separated in relatively pure form from a natural substance containing many admixed impurities.

**isomerization process** — although this term has a much broader meaning for a chemist, in the food field it refers to the enzymatic conversion of glucose to fructose in the manufacture of high fructose corn syrups.

**isomers** — compounds that can exist in more than one form, with different physical or chemical properties, even though they contain the same elements in the same proportions. There are two important types of isomerism: geometric and positional.

**Italian bread** — in the U.S., describes a loaf very similar to, indeed often indistiguishable from, French bread; hearth-baked loaves made from a lean dough containing little or no shortening, milk, or sugar.

**Ivarsson mixer** — in continuous breadmaking systems, one of the original types of dough developing machines.

**izyum** — *Russia* raisins.

## -J-

**jacket** — a double-wall construction feature of certain mixing and cooking vessels, allowing circulation of a heated or cooled liquid between the walls so as to adjust the temperature of the vessel's contents.
**jägerbrötchen** — *Germany* yeast-leavened bread rolls made from a blend of wheat flour with a relatively small proportion of rye flour.
**jaggery** — *India* crude cane sugar, unrefined.
**jaibas** — *Mexico* half-moon shaped turnovers made from a flaky dough, filled with jam or jelly.
**jaiphal** — *India* nutmeg.
**jaje** — *Serbia/Croatia* egg; plural is "jaja."
**jalea** — *Spain* jelly.
**jälkiuunileipä** — *Finland* rye bread baked a long time in a warm oven.
**jalousies** — *France* little puff pastry cakes.
**jam** — a preparation formed by cooking pieces of fruit (with or without seeds and usually without rinds) with a large amount of sugar, and sometimes with added pectin, until enough water is evaporated so that the residue forms a coarsely textured but soft gel.
**jam doughnut** — *UK* jelly doughnut; bismarck.
**jam roll** — *UK* jelly roll.
**japonica rice** — a subspecies of *Oryza sativa*, or common rice.
**jäst** — *Sweden* yeast.
**jästpulver** — *Sweden* baking powder.
**jaunes d'oeuf** — *France* egg yolks.
**javanica rice** — a race or subspecies of *Oryza sativa*, or common rice, grown principally in Indonesia.
**javitri** — *India* mace.
**jawar** — *India* barley.
**jeera** — *India* cumin.
**jelly** — a semi-solid food material prepared by boiling fruit juice, sugar (or other sweeteners), and sometimes pectin and acid. Most forms of jelly are covered by federal standards of identity.
**jelly crystals** — a dry granular mixture containing sugar, gelatin (or some other jellifying substance), colors, flavors, etc.; used for preparing imitation jellies and jams for fillings and the like.
**jelly roll** — a baked product made by spreading jelly (usually imitation) on a thin layer of cake, rolling the combination into a cylinder, and cutting pieces crosswise to show the internal spiral of filling.
**jelly wreath** — a rolled, flattish ring of basic sweet dough containing jelly filling; baked.
**jengibre** — *Spain* ginger.

**jet** — *France* bulge in a loaf resulting from slitting the top before baking.

**Jet-cut cookie former** — a cookie-dough shaping machine based on a rotating molding cylinder having a plunger in each die cavity for ejecting the dough blank; particularly useful with sticky and otherwise troublesome doughs.

**jordgubbstårta** — *Sweden* sponge cake finished with whipped cream and strawerries.

**jordnötter** — *Sweden* peanuts.

**jugo** — *Spain* fruit juice.

**julekage** — *Denmark* a rich, yeast-leavened coffee cake or stollen, made for Christmas.

**junior cereals** — foods consisting mostly of cereal products with additives for better nutritional balance, intended for children who have graduated from the infant versions of the same type of food, the improvement being primarily greater textural interest.

## -K-

**kääritortuu** — *Finland* Swiss roll.
**kabbouri** — *Egypt* semicircular loaf made from a mixture of corn and wheat flours, salt, and yeast; may include enriching ingredients such as honey or egg. The dough is given a short fermentation and proof, then baked to a relatively low moisture content.
**kab el ghzal** — *North Africa* almond-flavored, crescent-shaped cookies.
**kacamak** — *Serbia/Croatia* cornmeal mush.
**kadaifi pastry** — *ME* (various spellings) a threadlike pastry made by pouring a flour and water batter through a sieve onto a griddle and then quickly sweeping off the threads before they become brown. Can be rolled around fillings of various types. Looks a little like shredded wheat.
**kadin göbegi** — *Turkey* a kind of hot water dough, prepared similarly to cream puff dough and fried in shallow oil to give small golden-colored round pieces with a hole in the middle. Coated with syrup after cooking.
**kaffekage** — *Denmark* coffee cake.
**kafir** — white-hulled sorghum.
**kager** — *Denmark* cakes, cookies.
**kahvileipä** — *Finland* coffee cake, incl. cakes, sweet rolls, and pastries.
**kaiser roll** — a bread-type hard roll resembling from the top a sort of rosette with several curved depressions radiating from the center to the edge. Now made by stamping proofed dough pieces with an appropriately shaped cutter.
**kaiserschmarren** — *Germany* rich formula, fluffy pancakes containing raisins or raisin paste; served with a topping of fruit compote or chocolate sauce.
**kaiten-yaki** — *Japan* Japanese muffin containing bean jam.
**kaju** — *India* cashew nuts.
**kaka** — *Sweden* cake; cookie.
**kakao** — *Germany* cocoa.
**kakaod** — *Estonia* cocoa.
**kakku** — *Finland* cake.
**kalács** — *Hungary* cakes.
**kalakukko** — *Finland* pie made of whitefish and pork, baked in rye dough.
**kalamboki** — *Greece* corn.
**kale brose** — *Scotland* oatmeal and cabbage soup.
**kali mirch** — *India* black pepper.
**kandering** — *Sweden* icing.
**kanel** — *Sweden* cinnamon.
**kanela** — *Greece* cinnamon.
**kanelbulle** — *Sweden* cinnamon roll.

**kaneli** — *Finland* cinnamon.
**kao mao** — *Thailand* unhusked rice that has been flattened and roasted; used as a coating for fried foods (e.g., bananas).
**káposztás pogácsa** — *Hungary* biscuits containing fried cabbage.
**karask pärmiga** — *Estonia* barley bread.
**karaya gum** — a natural gum that swells to form a viscous gel when heated in water. Used to thicken and stabilize toppings and other bakery adjuncts.
**kardemumma** — *Sweden* cardamom.
**karidha** — *Greece* coconut.
**karidhia** — *Greece* walnuts.
**karidhopita** — *Greece* walnut cake.
**karidhopita athinaïki** — *Greece* walnut cake soaked with a cinnamon syrup.
**karikevma** — *Greece* spice.
**karintou** — *Japan* cookie prepared by frying (a) a yeast-fermented dough made from hard wheat flour or (b) a chemically-leavened dough made from soft wheat flour.
**karjalanpiirakka** — *Finland* a thin crisp pastry shell made of rye dough and filled with mashed potatoes or rice.
**karnemelk** — *Netherlands* buttermilk.
**kása** — *Hungary* porridge-like food made principally of grains such as millet, with flavoring and enriching ingredients such as meat.
**käsekuchen** — *Germany* cheesecake.
**kasha** — buckwheat, usually prepared by roasting or frying before boiling, and often used as an ingredient in soups, stews, etc. *Russia* gruel, porridge.
**kasha mannaya** — *Russia* farina porridge.
**kasha pshonnnaja** — *Russia* millet gruel.
**kasha pyerlovaya** — *Russia* pearl barley gruel.
**kasha risovaya** — *Russia* boiled rice.
**kas kas** — *Malaysia* poppyseed.
**kastanje** — *Netherlands* chestnut.
**kataïfi** — *Greece* a confection made from sugared noodle threads, almonds, walnuts, and syrup.
**kau fu** — *China* washed wheat gluten, resembling partially cooked dough, spongy in texture and appareance; added to braised dishes as a meat substitute.
**kaurakeksi** — *Finland* oat cracker.
**kaurapuuro** — *Finland* oatmeal.
**kawara-senbei** — *Japan* a tile-shaped cracknel made principally of wheat flour.
**käx** — *Sweden* cookie.

**kazandibi** — *Turkey* a pudding based on ground rice, flavored with rosewater.
**kedgeree** — *India* a savory rice dish containing a fairly large proportion of cooked flaked fish; highly seasoned with pepper, etc.
**keedetud koogid** — *Estonia* a kind of fried cookie made from a chemically leavened dough, usually flavored with lemon or vanilla.
**keik me karitha** — *Greece* coconut cake.
**keks** — *Serbia/Croatia Germany Sweden* cookies, sometimes crackers.
**keksi** — *Finland* cookie, cracker.
**kenyér** — *Hungary* bread.
**kerasia** — *Greece* cherries.
**kermakakku** — *Finland* sponge layer cake with cream and jam filling.
**kermaleivos** — *Finland* cream pastry.
**kermavaahto** — *Finland* whipped cream.
**kernel** — the grain of cereals, nuts, etc.; a large seed present in some fruits, such as peaches, apricots, and plums
**kernel paste** — apricot kernels ground to a very fine particle size with sugar and/or some other sweetener. Used in the same way as almond paste, as a texturizing and flavoring material. Very common ingredient of fillings for sweet goods such as coffee cakes. Although regarded by many as merely a cheaper substitute for almond paste, it does have an attractive flavor of its own.
**kærnemælk** — *Denmark* buttermilk.
**keskes** — *Arabian* perforated bowl (like a strainer) for holding couscous that is being steamed.
**khachapuri** — *Georgia* cheese pie served hot.
**khalvas** — *Greece* a sugary loaf made from farina and almonds.
**khameri roti** — also, "khamir." *India* To make this bread, a sponge composed of whole wheat flour, yogurt, salt and sugar is fermented overnight, then doughed-up with additional flour, baking soda, and water, and baked.
**khao tom** — or, "khao tom gung." *Thailand* congee.
**khejoor** — *India* dates.
**khilopittes** — *Greece* square noodles.
**khlyeb** — *Russia* bread.
**khubz** — *ME* a yeast-leavened loaf resembling pita bread.
**khvorost** — *Russia* crisp, flaky pastry, deep-fried.
**kifla** — *Serbia/Croatia* crescent roll.
**kiflice** — *Serbia/Croatia* crescent-shaped cookies, usually containing nuts.
**kiks** — *Denmark* cookie.
**kilning** — the removal of most of the water from a bed of germinated barley by passing hot air through the mass.
**kipfel** — also, "kipfere" and "kippels," *Germany* crescent-shaped almond-flavored biscuit.

**kirsch** — a liquor distilled from fermented cherry juice. Occasionally used as a flavoring for gourmet bakery foods, but more often found as an ingredient in continental confections.
**kirsebær** — *Denmark* cherry.
**kisela strudla** — also, "kiselo testo." *Serbia/Croatia* Similar to strudel, a rolled flaky pastry with a filling of crushed poppy seeds.
**kishmish** — *India* raisins.
**kisra** — *Sudan* flatbread made from a thick paste of sorghum flour, water, and sourdough that has been fermented about 14 hours. Baked for less than a minute on an oiled metal or clay surface. A minor amount of wheat or millet flour is often included in the dough.
**kiss** — meringue deposited as a kind of small mound, then baked slowly until it is almost dry. Generally, browning is to be avoided.
**kis sütemény** — *Hungary* teacakes or cookies.
**kitir** — *Serbia/Croatia* popcorn.
**kjeldahl method** — a procedure for determining total combined nitrogen in foods, basically by converting to ammonia and titrating the distillate. The nitrogen so determined can be multiplied by a factor, such as 5.7 for flour, which gives the approximate protein content of the material.
**klobasniky** — *Czechoslovakia* a small sausage baked in a bread roll.
**kloss** — *Germany* dumpling.
**kluski** — *Poland* noodles.
**klyotski** — *Russia* dumplings cooked in soup, often made principally of bread crumbs mixed with a minor proportion of meat.
**knäckebröd** — *Sweden* crisp bread, similar to Ry-Krisp.
**knækkebrod** — *Denmark* a quickbread in loaf form, often containing a large percentage of oatmeal; leavened with baking soda and sour milk.
**knafi** — *ME* see "kadaifi."
**knead** — to work a dough without cutting or tearing it, with the intent of developing the dough and/or releasing excess gas.
**knedliky** — *Czechoslovakia* dumplings
**knish** — a kind of large (perhaps 2 to 4 ounces) dumpling sometimes made of yeast-leavened dough, rolled out, filled with onion or other savory materials. Many variations are known, such as those made with a potato-based jacket. Some versions intended to be eaten out of hand.
**knock-back** — releasing excess gas from a fermenting mass of dough by any kind of manipulation or mechanical treatment that accomplishes this purpose.
**knödel** — *Germany* dumpling.
**knöpfli** — *Germany* thick noodle.
**knyte** — *Sweden* filled puff pastry similar to a turnover.
**koek** — *Netherlands* cake, including gingerbread.
**koekje** — *Netherlands* cookie.

**koh** — *Serbia/Croatia* porridge made from cracked wheat groats or cornmeal, with hazelnuts added.

**kohana fu** — a dried form of washed gluten in small pieces; used as a component in instant noodles and soups.

**kohupiima korp** — *Estonia* cheese cake.

**kokice** — *Serbia/Croatia* popcorn.

**kokojyväleipä** — *Finland* whole meal bread

**kokoskaka** — *Sweden* coconut macaroon.

**kokosnoot** — *Netherlands* coconut.

**kolaches** — also, "kolacz," "kolacky," etc. Many different formulas and procedures for making this pastry or cookie have been published; the prototype seems to have come from Czechoslovakia. A common present-day form is a small yeast-leavened sweet bun with a minor amount of jam or some other type of fruit filling centered on the top. Among the many variations are unleavened- and chemically-leavened pastries and cookies.

**kollane pühadesai** — *Estonia* Estonian Easter cake; generally a yeast-leavened dough, braided in strips, and flavored with saffron and orange peel.

**köömneid** — *Estonia* caraway seeds.

**kopenkhayi** — *Greece* sweet roll with almond filling, soaked in syrup.

**kornfleks** — *Serbia/Croatia* corn flakes.

**korngryn** — *Sweden* barley.

**korppu** — *Finland* rusk.

**korsan** — *Arabia* disk-shaped flatbread of about 2 oz piece weight made from whole wheat flour, water, and salt. Given two short fermentation periods and one proof period. Thin sheets baked almost to dryness.

**korvapuusti** — *Finland* cinnamon roll.

**kosher** — a food product conforming to Jewish dietary laws, i.e., ritually "clean."

**koulibiac** — *Russia* sauteed onions and salmon placed on a small sheet of brioche paste, then covered with chopped or sliced hard-boiled eggs with boiled rice, then the dough sealed around the filling and baked.

**kourambiethes** — *Greece* almond shortbread biscuits.

**kovrizhka** — *Russia* honey cake.

**kræmmerhus med flodeskum** — *Denmark* pastry cone filled with whipped cream and topped with jam.

**kräm** — *Sweden* (1) dairy cream. (2) Custard. (3) Thickened stewed fruit.

**kransekage** — *Denmark* a sort of pyramidal cake formed of almond macaroons.

**kransekagemasse** — *Denmark* marzipan.

**kranzkuchen** — *Germany* ring-shaped cake.

**krapfen** — *Germany* 1. Fritter. 2. Jelly doughnut.

**kreätopitta** — *Greece* pie with filling of seasoned ground meat.

**krema** — *Greece* cream.
**krem pita** — also, "krim torta." *Serbia/Croatia* napoleon.
**kremschnitte** — *Germany* a pastry confection resembling napoleons.
**krent** — *Netherlands* currant.
**kreplach** — noodle dough folded around fillings such as chopped liver; almost always cooked by boiling, as in soup.
**kringler** — *Denmark* pretzels, including sweet doughs in pretzel shape.
**krofne** — *Serbia/Croatia* doughnuts.
**krokant** — almond pieces mixed with molten caramelized sugar and, when cooled, broken into small pieces or ground. Used to flavor confections, etc.
**kruche ciasto** — *Poland* shortcake, often made of flour, powdered sugar, butter, egg yolks, and sour cream.
**kruchy** — *Poland* also "kruche," etc. A flaky-dough pastry, usually of the fat mix-in rather than roll-in type. Often finished as a fruit torte or pie.
**kruh** — *Serbia/Croatia* bread.
**krupa** — *Russia* gruel, porridge, grits.
**krupuk** — *Indonesia Malaysia* thin, dried wafers of cassava slices or tapioca starch or pastes from other plants; may be flavored with shrimp and the like; deep-fried to serve as a snack or garnish.
**krusbärspaj** — *Sweden* gooseberry tart or pie.
**krydda** — *Sweden* spice.
**kryddpeppar** — *Sweden* allspice.
**kryendyel** — *Russia* coffee cake, usually in a figure-8 shape.
**kserotigna** — *Greece* deep fried cookies. The dough, which is leavened with beaten eggs and flavored with cinnamon, anise, etc., is formed into strips that are twisted, tied, etc. After frying, draining, and cooling, the cookies are dipped in (or drizzled with) a honey syrup.
**kuchen** — *Germany* coffee cakes and the like made of yeast doughs.
**kugelhopf** — an Austrian sweet yeast cake containing raisins or currants and baked in a special high crownlike pan.
**kuglof** — *Serbia/Croatia* rich cake containing raisins and almonds and baked in a bundt-like pan, Similar to "kugelhopf."
**kukuruza** — *Russia* corn.
**kukuruz no brasno** — *Serbia/Croatia* cornmeal.
**kukurz** — *Germany* maize, Indian corn.
**kulchas** — *India* round, white-flour flatbread, deep-fried. Leavened by various means. Often split and filled in the same way as pita.
**kulich** — *Russia* tall cake containing raisins and often iced with marzipan, an Easter specialty.
**kulura** — *Greece* a kind of round bread of old-fashioned style.
**kulurakia** — *Greece* round cookies.
**kuluri** — *Greece* doughnut-shaped bread rolls sprinkled with sesame seeds.
**kumina** — *Finland* caraway.

**kümmel** — a liqueur flavored with cumin and caraway. Infrequently used to flavor bakery goods, but occasionally used in European confectionery. *Germany* Caraway (the spice).
**kurabiye** — *Turkey* shortbread cookies of butter, flour, and sugar.
**kurambies** — *Greece* almond cookies.
**kürbis** — *Germany* pumpkin.
**kuri-manju** — *Japan* chestnut bun with bean jam filling, the dough being chemically-leavened.
**kuvertbröd** — *Sweden Denmark* French roll.
**kuzina** — *Greece* (1) kitchen. (2) Cuisine, a tradition or style of cookery.
**kvasac** — *Serbia/Croatia* yeast.
**kyeks** — *Russia* plum cake.

## -L-

**label** — (1) Technically, a piece of paper or plastic in or on the container of a of an ingredient, retail product, etc. (2) Legally, any printed or graphic matter in or on the product or its container, or accompanying it in some other way, that purports to describe some quality of the product.

**labeling** — the printed information on the package or label or otherwise accompanying a product that describes some aspect of the product, including its effects, method of use, composition, etc.

**lace cookies** — thin and chewy cookies, with irregular holes that form during baking, usually chemically leavened and with high sugar content.

**lactalbumin** — also, "alpha-lactalbumin." One of the proteins that remain dissolved in whey when casein is precipitated from milk. It can be coagulated by heat and is used as a component of some fat replacers.

**lactase** — an enzyme that splits lactose into the monosaccharides galactose and glucose.

**lactated monoglycerides** — monoglycerides of fatty acids that have been reacted with lactic acid to form compunds, some of which have emulsifying properties.

**lactic acid** — the principal acid in sour dairy products; it is formed by certain bacteria acting on lactose and other sugars; it is a component of the flavor formed by rye sours.

**lactic acid bacteria** — microorganisms that form lactic acid and other flavoring materials in the fermentation process. They are important contributors to flavor in most types of sourdough breads.

**lactic acid rest** — in brewing, the holding of the mash at about 95°F for 30 to 60 minutes, to promote acidification early in the mashing cycle.

**lactobacillus** — any organism belonging to the genus of Gram-positive non-motile lactic-acid-producing microorganisms in the family Bacteriaceae; some species are important factors in sourdoughs.

**lactoglobulin** — also, "beta-lactoglobulin."a water-soluble protein found in milk and whey. When suitably modified, can be useful in fat replacers.

**lactometer** — a form of hydrometer or density bob calibrated to read in terms of the specific gravity of fluid milk products.

**lactose** — milk sugar; not fermented by yeast and has low sweetness. Chemically, a disaccharide containing glucose and galactose residues.

**ladhi** — *Greece* oil.

**Lady Baltimore cake** — rich white layer cake with fruit and nut filling and white icing.

**ladyfingers** — oblong sponge cakes typically about 1 inch wide, 3.5 inches long, and 0.3 inch thick. Most examples have a very light texture, are high in egg content, and are flavored with vanilla.

**lagkage** — *Denmark* layer cake, typically filled with either whipped cream, jam, custard, or some combination of these.

**lahmacun** — *Turkey* a yeast-leavened dough that has been made into a disc by pressing and rolling, then covered with a tomato paste concoction before being baked. Similar in many ways to pizza.

**lahvosh** — *Middle East* round and crisp cracker or bread made of wheat flour, malt, sesame seeds, and yeast (or sourdough); very thin.

**lake** — in food color terminology, a dye that has been reacted with ions of a metallic element (aluminum or calcium) so as to produce an essentially insoluble pigment; lakes are useful in many applications where a water soluble compound will not function satisfactorily, as in fat-based confectionery coatings.

**lame** — *France* a special knife or blade used to cut slits in dough pieces.

**laminating** — forming a structure of dough layers or of alternating layers of dough and shortening. An important step in making soda crackers, puff pastry, etc.

**laminators** — also, "laminaters." Machines for laminating dough; there are several different methods in use.

**lamingtons** — *Australia* small cubes of sponge cake that have been dipped into melted chocolate, then rolled in grated coconut.

**lanche** — *Portugal* snack.

**languette** — *France* the central part of a croissant.

**lap ovens** — industrial baking devices in which trays are carried back and forth within a heated chamber by a conveying system of chains, sprockets, and curved tracks.

**lapsha** — *Russia* noodles.

**lard** — rendered hog fat.

**lasagna** — *Italy* thick cooked pasta layers alternating with tomato, sausage meat, cheese, etc. and baked in the oven. Also, the pasta sheets sold for this purpose.

**laskiaispulla** — *Finland* roll filled with almond paste and whipped cream.

**latent heat** — heat taken up when a substance changes its state, as when water changes to steam or to ice without showing a change in temperature. Contrasted to sensible heat, which is heat that causes a change in temperature of the affected material.

**latke** — *Yiddish* pancake consisting mostly of shredded potatoes.

**latte** — *Italy* milk.

**lauric fats** — these ingredients typically contain 40 to 50% lauric acid in combination with lesser amounts of other relatively low molecular weight fatty acids. Lauric fats are obtained from the fruits of various types of oil palms, such as coconuts and palm kernels.

**lavash** — *Russia* unleavened white bread.

**layer cake** — although this designation would generally apply to any kind of cake baked in layers, the term is largely restricted in practice to high ratio cakes without nut or fruit ingredients that are iced or frosted whether in single or multiple layers. from small to large diameter.

**L-cysteine** — cysteine of the conformation found most widely in nature.

**leaf lard** — lard made from the fat taken from a hog's body cavity; it has the highest melting point (is the "hardest") of any of the unmodified lards.

**lean doughs** — doughs made primarily with flour, water, salt, yeast, and perhaps malt but with little or no enriching ingredients such as shortening, milk, and sugar.

**leaven** — originally, a portion of dough saved from day to day and used to inoculate a new batch of dough so it would ferment properly; now, the word is occasionally used as a short form of "leavening."

**leavener** — any substance used to generate gas inside a dough so as to provide the typical bubbly internal structure of bread, cakes, etc. Yeast, baking powder, and ammonium bicarbonate are the most common leaveners. Heat-generated water vapor and expanding air also contribute to the leavening of bakery products.

**leavening acids** — compounds added to any chemical leavening system (baking powder, batters, doughs) for the purpose of releasing carbon dioxide in gaseous form from carbonates or bicarbonates that have been added to the system. Cream of tartar, sodium aluminum sulfate, monocalcium phosphate, and sodium acid pyrophosphate are examples.

**leavening agents** — those substances or materials (including steam and air as well as yeast and baking powder) that cause an increase in volume of a dough or batter.

**lebkuchen** — *Germany* a dense, firm-textured European cake (or cookie) made with rye flour and honey, and highly spiced. Usually contains glazed fruits. Many variations exist, some of them characteristic of certain localities, such as Bremen pepper cake, Baseler lebkuchen, and Nürnburger lebkuchen.

**leche** — *Spain* milk

**lecithin** — chemically, applies to certain kinds of phosphatides occurring in both plants and animals, but the food ingredient "lecithin" is almost always obtained from crude soybean oil. Widely used in foods as an emulsifier and wetting agent.

**leckerli** — *Germany* honey and ginger-flavored biscuit.

**lefse** — *Sweden* flat bread made from potatoes or potato flour, wheat flour, water, salt, etc.

**legumbres** — *Spain* vegetables.

**lehetainas** — *Estonia* puff pastry.

**leipä** — *Finland* bread. The plural is "leivät."

**leite** — *Portugal* milk.

**leivitetty** — *Finland* breaded.

**lekach** — *Yiddish* a honey cake; lebkuchen.
**leivos** — *Finland* cake, pastry. The plural is "leivokset."
**lekvar** — dried prunes ground with about an equal quantity of water or corn syrup. Used as characterizing ingredient in fillings for Danish pastries, etc.
**lemon curd** — *UK* a kind of lemon custard, often made without milk. Sometimes used as a tart filling.
**lenja pita** — *Serbia/Croatia* apple pie.
**levadura en polvo** — *Spain* baking powder.
**levulose** — fructose.
**levure** — *France* yeast.
**levure en poudre** — *France* baking powder.
**light meal** — the ground-up scrap of light-colored and mildly flavored cookies that have been rejected or returned for some reason. Used at low levels as a bulking or non-characterizing ingredient in cookie doughs.
**lignin** — a mixture of substances physiologically related to cellulose and with it forming most of the structural part of woody tissue.
**lihapiirakka** — *Finland* pie filled with rice and chopped meat.
**lima** — *Spain Portugal* lime (the fruit).
**limao** — *Portugal* lemon.
**lime** — (1) A citrus fruit, typically greener, sourer, and smaller than lemons; Mexican, Persian, and Key are three commercial types of limes. Key lime pies are considered great delicacies although the flavor of most commercial examples is more closely connected with plants in New Jersey than with trees in Florida. (2) The mineral substances calcium oxide (unslaked lime), calcium hydroxide (slaked lime), or calcium carbonate (air-slaked lime).
**limón** — *Spain* lemon.
**limone** — *Italy* lemon.
**limpa** — *Sweden* rye bread.
**limppu** — *Finland* sweetened rye bread.
**LIMS** — laboratory information management system; a complex of computers, software, input devices, connections, etc., for processing and analyzing results, etc.
**limun** — *Serbia/Croatia* lemon.
**linguine** — *Italy* noodles in the form of flat, long, strips.
**linkage** — the specific arrangement by which atoms or combinations of atoms are joined together to form molecules.
**linse** — *Denmark* custard pastry.
**Lintner value** — a figure obtained by determining the rate at which malt can produce reducing sugars from soluble starch under defined conditions. Bread bakers find it an important specification for malt preparations.
**Linzer torte** — a rather thin, usually circular, layer of shortbread made with high levels of butter and filled or topped with raspberry jam.

**lipase** — an enzyme that breaks fats into free fatty acids and glycerol.
**lipid** — also, "lipide." Any of the group of fats and other esters that possess analogous properties.
**lipophilic** — having an affinity for lipids and similar substances.
**lipos** — *Greece* fat.
**lipoxidase** — an enzyme that oxidizes lipids and related compounds. As a baking ingredient, lipoxidase is used to oxidize the yellow pigment of flour, thereby lightening the color of the dough and the finished bread.
**lipoxygenase** — an enzyme that has been used to bleach bread doughs by oxidizing the yellow pigment carotene.
**liqueur** — a compound of ethanol (often in the form of raw brandy), sweeteners, flavors, colors, and other ingredients. Liqueurs can be used as flavors in bakery products, but are very expensive for this purpose.
**liquid sponge** — a sponge or pre-ferment containing flour, made with enough water so that the finished intermediate can be transferred, stored, and measured by liquid handling procedures. An essential part of continuous bread-making operations.
**liquid sugar** — a syrup particularly suited for bulk handling techniques, usually consisting of a 67% aqueous solution of beet or cane sugars; these syrups almost always contain a small amount of invert sugar.
**live bin** — a hopper provided with vibrating means for the purpose of improving uniformity of delivery of ingredients that are subject to bridging or flooding (e.g., flour, powdered sugar).
**loaf bread** — usually means non-sweet yeast-leavened bread baked in multi-portion units that are sliced or torn apart to yield individual servings.
**loaf cake** — cake batter that has been baked in bread pans or similar deep rectangular containers.
**loaf molders** — machines for performing the final shaping operations on the dough pieces that are to be made into pan bread.
**loaf volume** — the volume (usually expressed in cc) of a loaf of bread.
**lobe-type pump** — machine for moving fluids by a non-circular rotor moving inside a close-fitting chamber.
**locust bean gum** — carob gum, obtained from the seed of the locust bean tree. Used to increase viscosity and retain moisture in icings, fillings, and batters. Does not normally form gels.
**loempia** — *Netherlands* egg/spring roll of the Chinese pattern.
**log pretzels** — pretzels in the shape of thick, straight rods.
**lokma sade** — *Turkey* bite-sized fried pastries that have been dipped in a plain syrup.
**long-flake crusts** — pie crusts that exhibit distinct stratification, and when broken tend to shatter into large, relatively thin flakes.
**long-grain rice** — in addition to having kernels that are elongated, as compared to some other varieties of rice, these types usually exhibit a com-

paratively high amylose content and a moderately high gelatinizing temperature.

**long john** — a rectangular doughnut (no hole), perhaps 3 to 5 inches long, 2 to 3 inches wide, and an inch or so thick. Almost always made from yeast-leavened doughs. Usually iced on top (only), and frequently filled with a pudding or jelly type of material.

**long patent** — a patent flour that includes a relatively large percentage of the total flour produced by a mill, sometimes up to 95% of the total flour.

**long system** — a milling system involving a relatively large number of reduction processes; it leads to perhaps 15 different products.

**lönnsirap** — *Sweden* maple syrup.

**loss-in-weight bins** — ingredient bins or hoppers provided with means for measuring amounts removed from them, as when flour is delivered from a hopper into a mixer.

**loukoumades** — also, "loukoumathes." *Greece* Honey puffs; fried yeast-leavened dough in small pieces, dipped in hot honey immediately after frying.

**louver** — a slanted slat arranged (usually in multiples) to allow air to flow in or out of a chamber, as in a dryer; often adjustable to control the rate of air flow. "Louvre" is an incorrect spelling of this word.

**Lovibond color** — color of, e.g., oil measured by a technique utilizing a series of red and yellow glass slides as standards for visually matching the color of light passing through the oil sample. Colors so measured are usually reported in numerical values of red and yellow. The Lovibond scale and associated equipment is an official measuring device of the American Oil Chemists Society.

**Lowerator** — a conveying device for removing and stationing racks emerging from a final proofing room.

**low milling** — grinding grain with rollers set close together, a method in general use before the so-called New Process became popular. The aim was to produce as much flour as possible in one grinding step.

**L-sugars** — enantiomorphs of the common D-sugars, generally not utilized by the human body and so have been suggested for use as non-caloric sweeteners.

**lukier** — *Poland* icing.

**lukumadhes** — *Greece* deep-fried puffs of dough, crisp and light; served warm with honey syrup dressing.

**lutein** — hydroxylated xanthophylls, yellow pigments found in wheat endosperm and elsewhere.

**lye hominy** — see "hominy."

**lye peeling** — see "WURLD" process.

**lysine** — an amino acid essential for human nutrition.

## -M-

**määmmi** — *Finland* dessert pudding of malted rye and rye flour, flavored with orange rind and served cold with cream and sugar.

**maapähkinä** — *Finland* peanut.

**mabatt** — *Egypt* same as kabbouri, but the loaf is larger in diameter.

**macadamia nut** — a nut of medium size, with a very hard shell, obtained from the *Macadamia ternifolia* tree (or shrub) of Australian origin; the nutmeat is off-white in color, firm in texture, approximately spherical in shape, and mild in taste.

**maçapao** — *Portugal* almond macaroon.

**macaroni** — in consumer usage, tubular alimentary paste of moderately large size; in technical usage, almost any shape made of semolina paste.

**macaroon paste** — a combination of almond paste and kernel paste.

**macaroons** — these baked confections were originally made of ground or finely chopped almonds mixed with two parts of sugar and bound with egg white, but now the word is more likely to be applied a kind of coconut cookie that includes some flour.

**macarrao** — *Portugal* macaroni.

**macarrones** — *Spain* macaroni.

**maccheroni** — *Italy* macaroni.

**mace** — a spice that comes from the same tree as the nutmeg. Mace is the orange-colored fibrous material surrounding the kernel of the fruit of *Myristica fragrans*, the kernel being nutmeg. Generally, mace has a milder and more "rounded" flavor than nutmeg, but is used in many of the same kinds of products, including pie fillings.

**MacMichael viscometer** — a rotating-disc viscosimeter used to measure the viscosity of soft wheat flour mixed with dilute lactic acid. The results are often useful in evaluating the suitability of a flour for a particular application.

**madeleines** — *French* small sponge cakes baked in a scallop-shaped mold; usually coated with apricot or raspberry jam and then dusted with grated coconut. May be further decorated with glace fruit bits.

**mælk** — *Denmark* milk.

**mahlakas rulltort** — *Estonia* jelly roll.

**maicena** — *Mexico* corn flour or cornmeal.

**maida** — also, "maida flour." *India* a finely ground, white flour from wheat, used for breadmaking.

**Maillard reaction** — the so-called "non-enzymic browning reaction" resulting when amino acids and reducing sugars react at high temperatures; it produces poorly defined brown-colored compounds that sometimes have objectionable aroma and taste,

maintenance — upkeep of property, machinery and equipment; the practice or function of keeping equipment and other operating facilities in good operating order by scheduled examination,lubrication, etc. Not the same as cleaning or sanitation, although some of the functions may overlap.
mais — *Germany* corn.
maïskolf — *Netherlands* corn on the cob.
maissi — *Finland* corn (maize).
maito — *Finland* milk.
maiz — *Mexico* corn (the dry field corn).
maize — corn, Indian corn, the plant *Zea mays* and the grain harvested from it.
majs — *Sweden Denmark* corn (maize).
majsmjöl — *Sweden* cornmeal.
mak — *Russia* poppy seeds.
makagigi — *Poland* a confection of almonds cooked in honey and sugar, then cut into strips. Something like nut brittle.
makaroner — *Sweden* alimentary pastes, macaroni-type products.
makaroni — *Finland Serbia/Croatia* macaroni-type products.
makaronia — *Greece* macaroni.
makea — *Finland* sweet.
make-up — the manual or mechanical manipulations of dough pieces that are required to prepare from bulk dough the shape that will enter the oven.
make-up time — the duration of the period required to process the dough from the end of bulk fermentation to panning.
makkai — *India* corn.
makkai ki atta — *India* corn flour/meal.
makki ki roti — *India* a thin flatbread, crisp on outsides, fairly soft inside, made from finely ground cornmeal.
mákosgubó — *Hungary* a curious dessert prepared from small balls of rich yeast dough by first baking until slightly brown, then briefly stirring in boiling water, and finally heating in hot oil. Usually covered with poppy seeds and sugar.
makowiec — *Poland* yeast-raised dough made up in jelly-roll format, with a filling composed of poppyseed, honey, nuts, etc.
makron — *Sweden Denmark* macaroons.
malic acid — a crystallizable hydroxy diacid found in apples, grapes, etc., also called hydroxysuccinic acid. It is being made synthetically.
maloug — *Yemen* thin, disk-shaped flatbread made from wheat flour, water, salt, and yeast. Dough fermented about two hours, then shaped by pulling, and baked inside a clay oven.
malt — barley that has been allowed to sprout under controlled conditions, and then de-sprouted, dried or roasted, and ground. It is valuable as a

source of flavor, color, fermentables, and enzymes; it is the characterizing ingredient of beer wort. Wheat, and perhaps other grains, can be processed in somewhat the same way.

**malta** — *Spain* malt.

**maltase** — an enzyme found in yeast, and elsewhere, which converts maltose into glucose.

**malted milk** — a confection and flavoring material originally prepared by heating milk products and barley malt together, then vacuum drying. Now sometimes made by combining dried malt and dried skim milk with flour and other ingredients.

**maltodextrins** — produced from starch by one of the conversion processes used for making corn syrup, but the conditions are adjusted so as to yield molecules larger than sugars that are much less sweet and somewhat less soluble than, e.g., glucose. Can't be fermented by bakers' yeast.

**maltose** — a reducing disaccharide formed from the chemical union of two glucose units; not quite as sweet as sucrose. It is a reducing sugar. Can be fermented by bakers' yeast after an adaptation period.

**maltose value** — a measurement of the amylase activity of flour; the amount of reducing sugar produced in one hour from 10 gm flour held under standard conditions is reported as mg of maltose.

**malt syrup** — an ingredient prepared by extracting barley malt with water and evaporating the resulting solution. It is high in maltose and has enzyme activity depending on the heat treatment applied.

**manchet** — obsolete term referring to a rather thin round loaf of yeast-leavened plain white bread.

**mandel** — *Sweden Denmark Germany* almonds.

**mandelbiskvi** — *Sweden* almond cookie.

**mandelkager** — *Denmark* almond cookies.

**mandeln** — *Germany* almonds.

**mandoletti** — *Italy* a confection consisting of white nougat, fruit, and egg whites.

**mandorla** — *Italy* almond.

**manestra** — *Greece* noodles.

**mani** — *Spain* peanut.

**manju** — *Japan* generic term for Japanese buns with or without filling.

**mano** — *Mexico* a stone cylinder, generally less than a foot long, often made with tapered ends so that it resembles a spindle. Used in conjunction with a metate to grind nixtamal.

**mansikkakakku** — *Finland* sponge layer cake with strawberries and whipped cream.

**mansikkaleivos** — *Finland* strawberry filled pastry.

**mansikkatorttu** — *Finland* strawberry flan.

**manteca** — *Spain* lard or other cooking fat.

**mantecadas** — *Spain* small cakes of the kind sold in frilled papers.
**manteli** — *Finland* almond.
**mantequilla** — *Spain* butter.
**mantequilla de cacahuates** — *Spain* peanut butter.
**manzana** — *Spain* apple.
**MAP** — modified atmosphere packaging; a method of increasing the shelf life of packaged foods by controlling the composition of the gases in contact with the product.
**maple syrup** — sap of the sugar maple tree that has been concentrated by heat evaporation. A flavorful but expensive sweetener.
**maräng** — *Sweden* meringue.
**maraschino cherries** — artificially colored and flavored cherries. Commercial goods of this name are no longer flavored with maraschino liqueur, at least not in the U.S.
**marble cake** — cake in which two or three differently colored cake batters have been combined in swirls before baking.
**marenki** — *Finland* meringue.
**margarine** — a plastic or flowable emulsion serving as a substitute for butter and containing a minimum of 80% fat. The non-fatty portion consists of some combination of water, milk products, salt, color, emulsifiers, flavor, and other additives.
**marille** — *Germany* apricot.
**maritozzo** — *Italy* soft roll.
**mark yang tong** — *China* so-called "malt sugar"; in form, a thick, molasses-like ingredient, often sold in small plastic tubs and used as a brush-on barbecue sauce; made by evaporating barley malt.
**marmeladha** — *Greece* jam.
**marmellata** — *Italy* jam.
**marone** — *Germany* chestnut.
**marranitos** — *Mexico* a moderately sweet, chemically leavened, baked cookie or cake, usually about 1 or 2 oz in weight, flavored with molasses (and, usually, with spices such as ginger), made in the form of a pig (hence, the name). Dark brown in color, usually not crisp.
**marrone** — *Italy* chestnut.
**marrons** — large sweet chestnuts of a special cultivated type, available as a puree, candied, or dried. Common as a flavoring and texturizing ingredient in Europe, but rare in the US.
**marrons glacées** — candied sweet chestnuts.
**marshmallow** — a soft confection foam made of whipped sugar syrup and corn syrup stabilized with gelatin and other gums and, usually, flavored with vanilla.
**marsipan** — *Sweden* marzipan; almond paste.
**massa** — *Portugal* (1) Dough. (2) Pasta.

**marzipan** — almond paste mixed with powdered sugar and corn syrup; traditionally, it contained some egg whites and was flavored with rosewater or orange-flower water. The term is also applied (at least in the U.S.) to fruit, vegetable, and flower shapes formed from colored marzipan and either consumed as separate confections or put on baked products as decorations.

**masa** — *Mexico* the raw material for tortillas and similar products; made by cooking field-corn kernels in a slightly alkaline solution, washing (sometimes), and grinding to a paste. More generally, any kind of dough.

**masa harina** — a dry mix from which tortilla dough can be prepared merely by blending with water.

**masa trigo** — *Mexico* a dry mixture for making the wheat flour type of tortilla.

**mashing** — in brewing, the programmed heating of a mixture of malt, water, and other materials for the purpose of extracting the maximum amount of desirable fermentables.

**mashing off** — in beer-making, a rest during which the temperature is typically held in the range of 167°F to 176°F for 5 to 10 minutes so as to inactivate most of the enzymes.

**masking** — covering a baked product, esp. a cake, with icing or frosting.

**maslac** — *Serbia/Croatia* butter.

**masoor dal** — *India* lentils.

**mast** — *Serbia/Croatia* lard.

**materials handling** — the physical procedures involved in receiving, transporting within the plant, and dispensing ingredients and other processing materials, packaging components, etc.

**matsusake fu** — *Japan* a piece of baked gluten, shaped and flavored to resemble mushroom slices.

**maturing** — allowing flour to age to improve its processing qualities, or adding chemicals to accomplish the same results.

**matzos** — also, "matzoth." A thin, unleavened bread acceptable for consuming during the Jewish Passover. Usually made from a low absorption flour and water dough developed by repeated sheeting and laminating before being docked and baked in a hot oven for about a minute. Has acquired some popularity as a snack food, leading to a proliferation of flavors and other modifications.

**matzo balls** — crushed or ground matzo mixed with chicken fat, eggs, and seasonings, and formed into small balls. Usually cooked in chicken soup.

**mazapán** — *Spain* marzipan, almond paste.

**mazurek** — *Poland* cakes or tortes made with flour, whipped egg whites, a high proportion of ground almonds, and other ingredients. Some recipes include mashed hard-cooked egg yolks. Found in many flavors and forms but usually flat and rectangular.

**mazurka** — *Poland* a cake made from relatively thin layers of baked mazurek dough separated by fillings of preserves, peanut butter, chocolate, dried fruits, etc. The top layer is generally frosted.
**meal** — coarsely ground grain.
**mealy crust** — a pie crust exhibiting very little flakiness due to uniform distribution of fat within the dough. On breaking, yields crumbs rather than flakes.
**mechanical bench** — a conveyor belt fitted with appliances and utensils for making coffee cakes, cinnamon rolls, and the like from raw dough pieces and adjuncts.
**med** — *Serbia/Croatia* honey.
**medium chain triglycerides (MCT)** — triglycerides (fats) containing C6, C8, and C10 saturated fatty acids. They are absorbed quickly by the body and are transported via the portal system. Conventional fats and oils containing C16 and C18 fatty acids are absorbed more slowly and are transported principally by the lymphatic system.
**mee** — *Malaysia* a form of egg noodles.
**mehl** — *Germany* wheat flour.
**mehlnockerl** — *Germany* small dumpling.
**mehlsuppe** — *Germany* soup made mostly of browned flour and broth.
**mein** — also, "mie." *Asia* noodles made from water and wheat flour or meal.
**mein jin pau** — also, "mianjin." *China* Golfball-size pieces of golden gluten, deep-fried and then used as an ingredient in soups, etc.
**mejorana** — *Spain* margarine.
**mel** — *Portugal* honey.
**mela** — *Italy* apple.
**melass** — *Sweden* molasses.
**melassa** — *Italy* molasses.
**mélasse** — *France* molasses.
**melasse** — *Germany* molasses.
**melaza** — *Spain* also, "melote." molasses.
**melba toast** — thin slices of bread toasted slowly to dryness and until they assume a uniform light brown color.
**meli** — *Greece* honey.
**melk** — *Netherlands* milk.
**melocotón** — *Mexico* peach.
**melokaridho** — *Greece* small cake made of honey and walnuts.
**melomakarona** — *Greece* another kind of pastry steeped in honey syrup.
**melting point** — the temperature at which a solid becomes a liquid.
**Melton Mowbray** — *UK* a meat pie, usually of individual size, intended to be eaten cold and containing seasoned chopped pork and anonymous materials in a gel.
**menagerie cookies** — essentially the same as animal crackers.

**menrui** — *Japan* noodles (generic).
**merengue** — *Spain Portugal* meringue.
**merienda** — *Spain* snack.
**meringe** — *Germany* also, "meringel." Meringue.
**meringue** — egg whites beaten with sugar; usually a small amount of acid (such as cream of tartar or lemon juice) is added to improve volume, stability, and color. If a flavor is added, it is usually vanilla.
**meringues** — shapes formed from meringue and then baked almost to dryness; served either as separate confections or used as a base for other fillings, toppings, etc.
**mermelada** — *Spain* fruit jam or preserves.
**mesh** — the number of wires (or threads) per inch in a sieve; can be expressed in different systems of measurements.
**metabolism** — the complex of chemical reactions that make up the life processes in organisms.
**metallized film** — plastic film that has been covered with a very thin layer of some metal (usually aluminum) by vapor deposition or other methods.
**metate** — *Mexico* a three- or four-legged stone utensil with a horizontal, flat, sloping surface; used with a mano for grinding nixtamal in making tortillas. Generally, the grinding surface is somewhat concave.
**methionine** — a sulfur-containing amino acid found in certain proteins, such as casein and egg albumin. Also made synthetically and used for nutritional supplementation of foods.
**methylcellulose** — a cellulose gum in which many of the hydroxyl groups of cellulose have been replaced by methoxyl groups. This treatment normally leads to a "stronger" gum.
**metric system** — a decimal (i.e., scaled by tens) system of weights and measures, with the meter and gram as bases of length and mass, respectively.
**mett** — *Estonia* honey.
**microcrystalline cellulose** — a gum derived from cellulose by chemical treatment, useful in foods as a texture modifier.
**micronutrient** — a trace element or an organic compound, such as a vitamin, that must be present in the diet to sustain health and growth.
**microorganisms** — in general, independently living organisms that are too small to be seen with the naked eye; more specifically, yeasts, bacteria, molds, and the like.
**microwave** — electromagnetic radiation of very short wavelengths, usually regarded as radiation between 1 cm and 100 cm in wavelength.
**microwave oven** — an oven that relies on the transformation of radiant energy of specific wavelengths to heat within the product to be cooked.
**microwave proofing** — use of microwave heating to accelerate the proofing process in yeast-leavened doughs.

**middlings** — (1) The larger particles coming from the floury part (endosperm) of the wheat berry during milling; some small bits of bran may also be present. (2) In the animal feed trade, "middlings" means fairly high grades of milling by-products, i.e., mixtures of bran, germ, some of the large endosperm chunks, etc.

**middlings rolls** — in a flour mill, roll stands the operation of which produces mainly middlings.

**mie** — *France* the crumb of bread, i.e., everything but the crust.

**miel** — *France Spain* honey.

**miele** — *Italy* honey.

**miga** — *Spain* the crumb of bread, i.e., the part that is not crust.

**miki** — *Philippines* egg noodles.

**mil** — one thousandth of an inch, a unit of measurement much used in the packaging film industry.

**milch** — *Germany* milk.

**mildew** — growths of fungi (perhaps other organisms as well) on food products and other materials, leading to the production of white areas and to other defects.

**mil-folhas** — *Portugal* flaky pastry sheets layered with a creamy filling after baking, similar to napoleons.

**milho doce** — *Portugal* sweet corn.

**milk** — unless modified, the term means "the lacteal secretion, practically free from colostrum, obtained by the complete milking of one or more healthy cows."

**milk bread** — white bread in which there is an amount of milk solids specified by federal standards.

**milk chocolate** — chocolate liquor to which has been added milk solids, sugar, flavors, and emulsifiers; must have a minimum chocolate liquor content of 10%.

**milk protein concentrates** — various commercial preparations based on milk from which most of the lactose and mineral solids have been removed; in some cases with increased casein content, in other cases with increased whey proteins content.

**milk replacers** — dried products intended to replace dehydrated skim milk at a lower cost, and consisting of some combination of caseinates, whey, modified whey, soy protein, corn syrup solids, and additives used for improving drying properties or functional characteristics.

**milk solids not fat** — dried skim milk.

**milk sugar** — lactose, q.v.

**millefeuille** — also, "mille feuille.' *'France* Puff pastry in general or napoleons in particular.

**millefoglie** — *Italy* puff-pastry and custard combination, similar in a general way to napoleons.

**millet** — a name that has been applied to a wide variety of cereal grains, even sorghum, but is more properly used for two tribes of the grass family, the Chlorideae and the Paniceae. The former includes African ragi (finger millet) as the only species of economic importance, while the latter tribe includes several species grown for food and feed in various parts of the world. *Panicum millaceum*, proso or common millet, is probably the only species grown to any extent in the U.S.

**millfeed** — a milling by-product, such as bran; any product of a mill not identifiable as flour.

**milling predictors** — results of tests or evaluations performed on grain that are correlated to the performance (e.g., percentage yield of flour) of the grain when it is passed through a given mill.

**milling quality** — a term of broad and rather indistinct meaning summarizing the many factors affecting a grain's response to milling processes.

**milo** — sorghum. *Greece* Apple.

**milopitta** — *Greece* apple tart.

**milpa** — *Spain* cornfield.

**mincemeat** — a filling for pies consisting of a cooked mixture of raisins, chopped apples, candied fruits, spices, beef suet, etc. At one time, mincemeat contained a fairly large percentage of finely chopped beef, but commercial products no longer include this ingredient..

**mineral oils** — hydrocarbons derived from petroleum; used for lubrication purposes in the bakery and, sometimes, as part of the pan grease mixture.

**minerals** — in foods, inorganic substances and particularly, the ash remaining after the organic materials have been burned off; also, the metallic elements essential in human nutrition, such as sodium, potassium, calcium, iron, manganese, etc.

**minjoni** — *Serbia/Croatia* small cakes filled with cream and iced.

**mint** — flavoring materials derived from the leaves of Menthus plants. Commercially useful varieties are spearmint, Scotch-type spearmint, applemint, pennyroyal, and peppermint. Synthetic oils chemically identical to the natural essences are also available. Mint is a rather uncommon flavor in bakery foods, but it is sometimes used in the fillings for sandwich cookies or as a supplementary flavor in chocolate coatings.

**miraculin** — an intensely sweet material extracted from the berry of a tree, *Sideroxylon dulcifieum*, found in tropical Africa.

**miso** — *Japan* a soft paste made from steamed soybeans that have been allowed to ferment after the addition of a starter made from rice and wheat (or sometimes other grains). Dozens, perhaps hundreds, of different kinds are known. Used as a flavoring ingredient for soups and many other types of foods.

**miswa** — *Philippines* wheat-flour noodles, usually sold in skeins.

**mitt** — an insulated glove or mitten used for handling hot objects.

## GLOSSARY OF CEREAL TECHNOLOGY TERMS 125

**mixer** — any device or machine intended for use in creating a more or less uniform blend of two or more ingredients; in processing doughs, mixers have the additional function of "developing" the material.

**mixing bowl** — the container affixed (permanently or temporarily) to a mechanical mixer for holding the ingredients as they are being stirred; also, a large hemispherical bowl used for manual blending of ingredients.

**mixing tolerance** — the relative capacity of a dough to withstand changes in mixing conditions, especially variations in mixing times.

**mixing tolerance index** — a figure determined by measurements on the curve drawn by a Brabender farinograph. Larger figures indicate the likelihood of greater mixing tolerance during production.

**mjölk** — *Sweden* milk.

**mleko** — *Serbia/Croatia* milk.

**mlinci** — a thin flatbread made in Croatia; similar to chapatis in preparation method.

**mocha** — originally, a special kind of coffee bean. Now, usually means a blend of chocolate and coffee flavors.

**mochi** — *Japan* a cake or patty made from steamed glutinous rice that has been pounded to a paste; various shapes and flavors.

**mochi gome** — *Japan* a glutinous rice having very small, oval, rather flat grains.

**modification** — in malting, the changes that occur in the barley kernel during the steeping and germination steps, generally related to the many enzyme reactions involved in sprouting.

**modulating burners** — gas or oil burners for ovens, fryers, etc., that adjust the heat intensity by controlling the rate of flow of fuel to the flames; i.e., the control is not limited to all-on and all-off.

**mohinga** — *Burma* cooked rice-vermicelli in a thin, creamy seafood curry, coconut flavored.

**moh loung ye baw** — *Burma* small flour and coconut dumplings filled with coarse sugar; served in a coconut milk sauce.

**moisture and volatile matter content** — determined as the weight lost by a food material after it has been heated for a prescribed time under controlled conditions. Reported as percent of the original weight of the sample.

**moisture vapor transmission rate** — (MVTR) the rate at which water vapor passes through a packaging film or other material, as determined under controlled conditions.

**Mojonnier test** — a method for determining the percentage of fat in dairy foods involving the ether extraction of butterfat from a sample, separating the ether solution, evaporating the solvent, and weighing the dry fat.

**molasses** — as a bakery ingredient, refers either to a syrup obtained as a by-product in the refining of cane sugar or made as the prime product by evaporating sugar cane syrup. Colors range from light tan to almost black, with flavor (acidity, bitterness, metallic) becoming stronger as the color becomes darker. Used as a coloring and flavoring ingredient in many foods.

**molasses cookie** — a category of chewy, brownish, usually spiced cookies of various sizes, containing a fairly high proportion of molasses. Chemically leavened and sometimes containing adjuncts such as nuts or raisins.

**molcajete** — *Mexico* a mortar shaped grinding utensil consisting of a hemispherical hollow in a stone base, used with a tejolote to grind nixtamal into the masa used for tortillas.

**mold** — (1) Hollow forms of plastic, metal, or (rarely) ceramics used to shape confections; if used for baking, they are commonly referred to as "pans." (2) A fungus, i.e., a mycelial microorganism.

**molder** — also, "moulder." A machine that shapes dough pieces for baking, as a bread loaf molder.

**molding aprons** — also, "molder aprons." An endless canvas belt of special weave that passes under a cookie molding drum and removes the dough blanks from the die cavities.

**mold inhibitor** — a chemical that can be added to a formula to delay fungal spoilage; also, can be sprayed on the surface of, e.g., English muffins.

**molletes** — *Spain* a circular sweet bun with colored paste forming a circle with four rays on top.

**monelin** — a natural sweetener of high intensity.

**mon-le-bway** — *Burma* fried rice flour batter made into crisp, thin sheets that are very light due to the presence of innumerable air bubbles.

**monocalcium phosphate** — also called MCP, calcium acid phosphate, and acid calcium phosphate. A mineral substance used as a leavening acid, yeast nutrient, and dough acidulant. Two forms of very different reaction rates in batters: the anhydrous form and the monohydrate.

**monoglyceride** — a chemical compound formed by the combination of one fatty acid unit with one glycerol unit. Used as surfactants and to delay texture staling of bread.

**monorail proofer** — a conveyor consisting of an elevated single rail serving as the track for wheels from which are suspended racks containing the fermenting dough products.

**monosaccharide** — a simple sugar containing 3 to 9 carbon atoms (6 carbons in all ingredient monosacchrides) with the same number of oxygen atoms and twice that number of hydrogen atoms; more complex sugars are made up of combinations of monosaccharides.

**monosodium glutamate** — an essentially flavorless compound that acts as a flavor enhancer or appetite stimulator.

**moongfalli** — *India* peanuts.

**moscarpone cheese** — *Italy* a soft, mild-flavored cheese, white or ivory colored; very high in butter fat and similar to cream cheese in other respects. Used as an ingredient in spreads, icings, etc.

## GLOSSARY OF CEREAL TECHNOLOGY TERMS 127

**mousse** — originally, a frozen dessert of sweetened and flavored whipped cream with fruit pieces, often with gelatine as a stiffener; the term is now frequently used for a whipped dessert high in fat, flavored (most often with chocolate) but not always frozen. Sometimes, even used for whipped starch puddings.

**MPN** — most probable number of *Escherichia coli*; the bacterial density, which, if it had actually been present in the sample under observation, would more frequently than any other [density] have given the observed analytical results.

**mqdrzki** — *Poland* fried cookies or pastries made of pot cheese, eggs, sugar, and flour. Aerated only by beaten eggs.

**muffins** — chemically leavened batters baked in the small cups of muffin pans, usually fairly sweet. Also, yeast-leavened soft doughs cooked in rings on a hearth, e.g., English muffins.

**muffin splitters** — rotating knives or fork-shaped implements that separate or partially separate the upper and lower halves of a baked English muffin.

**muka** — *Russia* flour.

**multigrain breads** — breads that contain substantial amounts of other grains in addition to wheat flour (or ground whole wheat).

**multi-tier proofer** — large final proofers in which each tray, after it is loaded, is raised individually to the top tier of the conveyor system.

**muna** — *Estonia Finland* eggs.

**munakook** — *Estonia* sponge cake.

**munariisipasteija** — *Finland* pasty with egg and rice filling.

**munavalgekoogid** — *Estonia* meringue cookies made with egg whites, sugar, and finely chopped almonds.

**munk** — *Sweden* doughnut.

**munkii** — *Finland* jelly doughnut.

**murot** — *Finland* breakfast cereals.

**mush** — cereal flour or meal made into a thick porridge or gruel with hot water; generally refers to corn meal mush in the US. Also refers to the same material after it has congealed upon cooling, usable then as a sliced piece for frying.

**mushi-manju** — *Japan* a steamed bun, chemically leavened and containing a bean jam filling.

**mushimono** — *Japan* steamed food (generic term).

**muskat** — *Germany* nutmeg.

**muskot** — *Sweden* nutmeg.

**mustokulura** — *Greece* grape cookies.

**mycotoxin** — a poison elaborated by fungi.

**mylar** — a polyester resin used to make films for food packages.

## -N-

**näkkileipä** — *Finland* crisp bread.
**nalesniki** — *Poland* crepes, pancakes, blintzes.
**nalistuki** — *Russia* crisp crepes filled with cheese.
**nama fu** — *Japan* fairly small chunks of moist gluten, vended as a food ingredient.
**nan** — *India* (also, "naan") a flatbread made from flour, yeast, sugar, salt, water, yogurt, and shortening. Although nan is usually yeast-leavened, chemically leavened versions are known. Similar in size and shape to chapatis.
**Naples biscuits** — originally, a thick (up to one inch) cookie made from a rich unleavened dough, and flavored with, e.g., rosewater, wine, and spices. In modern terms, any cookie to which the name "Naples" can be applied. .—
**napoleons** — sheets of baked puff pastry alternating with layers of cream filling, cut into rectangles and sometimes iced.
**nappage** — *France* colored and flavored transparent glazes applied to the tops of cakes and tarts to improve the food's appearance and shelf-life.
**naranja** — *Spain* orange (the fruit).
**nariyal** — *India* coconut.
**nasi** — *Malaysia* cooked rice.
**nasi goreng** — *Indonesia* a savory preparation based on a fried mixture of boiled rice and various meat, fish, and egg items. Coconut milk is added and other condiments and decorative bits applied.
**naspati** — *India* pear.
**Nassau tart** — *Old English* a baked pie or tart shell filled with a heated mixture of marmalade, butter, and eggs, then baked until the filling is puffed and browned on top.
**nata** — *Spain Portugal* cream, whipped cream.
**nata batida** — *Spain Portugal* whipped cream.
**natillas** — *Spain* custard.
**neige** — *France* preceded by "à la" or "en," indicates a product containing, as a separate component, beaten egg whites/meringue.
**nelke** — *Germany* clove.
**neohesperidin** — non-caloric high-intensity sweeteners made by modifying compounds obtained from the rinds of citrus fruits, especially grapefruit.
**neosugar** — a synthesized, low- or no-calorie sweetener.
**nesselrode pie** — rum-flavored Bavarian cream pie filling mixed with assorted preserved fruits and placed in a pre-baked pie crust; traditionally topped with chocolate curls.
**nests** — see "nidi."

## GLOSSARY OF CEREAL TECHNOLOGY TERMS 129

**neufchâtel cheese** — a soft cheese somewhat similar to cream cheese but lower in fat and therefore cheaper and lower in calories. Has been used as a replacement for cream cheese in some pastry fillings.

**neutral detergent fiber** — one form of food fiber, perhaps more accurately defined as one method of determining certain types of food fiber; gives higher results than the traditional crude fiber determination.

**neutralizing value** — a number describing the strength or power of a leavening acid, being the number of pounds of sodium bicarbonate required to neutralize 100 pounds of the acid.

**New Process** — the process of high grinding with millstones, using one or more lower regrinds, that came into use in the U.S. just before the introduction of modern roller milling. In comparison with very long continental New Process systems, the versions favored in the U. S. were sometimes described as "half-high grinding."

**N-flate** — trademark for a blend of materials such as emulsifiers, offered as an improver for cake batters.

**niacin** — nicotinic acid, a water soluble (B) vitamin; one of the nutrients required to be added to enriched flour; originally called "P. P.," or pellagra-preventive factor.

**nicotinamide** — the amide of niacin.

**nidel** — *Germany* also, "nidle." Cream.

**nidi** — raw noodles arranged in a shape somewhat resembling a bird's nest, then dried.

**nimbu** — *India* lime.

**nip** — as applied to a pair of mill rolls, the minimum distance between the corrugations of the cylinders as they rotate.

**nitrogen-blanketing** — the principle or practice of restricting access of food oils to oxygen (thus retarding the development of oxidative rancidity) by storing the oils under an atmosphere of nearly 100% nitrogen.

**nixtamal** — the soaked, cooked, unground corn kernels used to make masa. An intermediate in the preparation of tortillas.

**NMR spectrometers** — analytical instruments relying on the principles of nuclear magnetic resonance to determine quantities.

**nocciola** — *Italy* hazelnut.

**noce** — *Italy* nut.

**noce de cocco** — *Italy* coconut.

**noce de moscata** — *Italy* nutmeg.

**nockerl** — *Germany* small dumpling.

**nod** — *Denmark* nut.

**noix** — *France* nut.

**noix de coco** — *France* coconut.

**noix muscade** — *France* nutmeg.

**noklice** — *Serbia/Croatia* dumplings, noodles.

**nondiastatic malt** — malt syrup (wet or dried) that has been heat treated to such an extent that very little amylolytic activity remains.

**nonenzymatic browning** — the development of brownish colors (and, often, off-flavors) in food ingredients and products due to the Maillard reaction and similar reactions not mediated by enzymatic processes.

**nonfat dry milk** — also, "NFDM." Dried skim milk.

**non-nutritive sweetener** — any substance (such as saccharin) that can be added to foods to make them sweeter but which does not contribute calories to the diet.

**nontropical sprue** — a synonym for "celiac disease."

**noodle nests** — a layer of boiled noodles fitted between two metal sieves or "baskets" (roughly hemispherical) and deep-fried until crisp. Used, esp. in Chinese cuisine, as containers for stir-fried mixtures and the like.

**noodles** — a pasta made from a dough consisting of semolina or flour mixed with water, eggs, and salt. In the US, the egg ingredient is mandatory. There are many shapes, but all are basically rectangular, often fairly long thin strips. Noodles are boiled in water or stock, then added to various kinds of flavorings, condiments, meats, etc.

**noot** — *Netherlands* nut.

**no-time dough** — a straight dough, which through the use of more fermenting agents and higher temperatures than normal, and usually with the aid of more mechanical development in the form of mixing, has its fermentation period reduced from hours to less than about 20 min. These doughs are sent to make-up immediately after mixing, with a generally unregulated floor time during which some fermentation occurs.

**nötter** — *Sweden* nuts.

**nougat** — originally, candy made by stirring nuts (such as almonds or pistachios) into molten sugar, then adding a fatty ingredient to soften the texture. Now, usually, a slightly aerated mixture of sugar syrup, egg white, and condensed milk. Other mixtures are also inexactly called nougat.

**nouilles** — *France* noodles.

**noz** — *Portugal* nut, walnut.

**noz moscada** — *Portugal* nutmeg.

**nudel** — *Germany* noodle.

**nudle** — *Serbia/Croatia* noodles.

**nueces de acachù** — *Mexico* cashew nuts.

**nueces de Brasil** — *Mexico* Brazil nuts.

**nueces de Castilla** — *Mexico* walnuts.

**nuez** — *Spain* walnut; *Mexico* pecan.

**nuez moscada** — *Spain* nutmeg.

**nugget pretzels** — log pretzels of short length.

**Nulomoline** — a trade name for a standardized invert sugar syrup.

**nuoc mau** — *Vietnam* caramelized sugar used to flavor and color foods.

**nutating meters** — devices for measuring the flow of water or other liquids, depending for their action on the back and forth movement of a mechanism as alternate sides of a chamber are repeatedly filled and emptied..

**nut butters** — ground nuts (sometimes with minor additives) of fine particle size and smooth consistency; peanut butter is the prime example.

**nut cluster formers** — machines that form clumps of two or more nuts or nut pieces held together by, e.g., caramel or chocolate.

**nutmeg** — a spice frequently used in bakery products, made from the ground or grated kernel of the fruit of an East Indian tree, the same plant that produces mace.

**nutrient** — substance in food or drink that can be taken up by the body and used as a metabolite.

**nutritional supplementation** — the addition to a food product of essential nutrients for the purpose of making the food more suitable for specific dietary purposes.

## -O-

**Oakes mixer** — a popular type of continuous mixer for batters, marshmallow, etc., based on the principle of toothed or serrated disks rotating in a close-fitting circular chamber that can be pressurized.

**oatcake** — *UK* oat cracker.

**oatmeal** — various chopped or ground forms of the grain of the oat plant, sometimes lightly toasted, cooked as a hot cereal and used as an ingredient in baked products and ready-to-eat cereals.

**oats** — the seed of the cereal plant *Avena sativa*.

**obst** — *Germany* fruit.

**odrajahu** — *Estonia* barley flour.

**oeuf** — *France* egg.

**offal** — *UK* the millfeed fractions coming from a flour mill.

**ohrasämpylä** — *Finland* barley roll.

**ohukaiset** — *Finland* small thin pancakes.

**oil** — (1) In food processing, a natural or processed edible fat (triglyceride) that is liquid under normal storage or usage conditions. (2) In petroleum product nomenclature, hydrocarbon mixtures liquid at room temperature.

**oiled dough** — dough pieces that have had oil applied to their surfaces, generally just before baking; the usual purpose is to eliminate the need for greasing the pans, although other benefits are sometimes obtained,

**oil-fired oven** — an oven that derives its heat from the combustion of atomized fuel oil.

**oil seeds** — any plant seeds from which it is commercially possible to extract edible oil; soybeans and cottonseeds are examples.

**ojo** — *Mexico* pastries made of an outer circular layer of sweet dough with a central depression containing cake batter, usually vanilla flavored.

**o-kashi** — *Japan* cakes made of mashed (sweet) potato or mashed beans cooked with sugar, isinglass, seaweed, and sometimes sake; molded into blocks or fanciful designs. Not leavened.

**oklablomos** — *Greece* a round flatbread.

**okruglice** — *Serbia/Croatia* meat dumplings.

**ol** — *Germany* oil.

**oladi** — *Russia* fritter made from a yeast leavened dough, usuually served with jam, honey, or sour cream.

**old dough** — yeast dough that is sour and weak from overfermentation, having been held too long or at too high a temperature; will give dark, sour bread, the loaves exhibiting low volume, boldness, rough pan crust, dull top crust, and irregular collapsed cell structures.

**oleo** — also, "oleo oil." The liquid fraction of edible beef fat, rendered from tallow, and the like.

**óleo** — *Portugal* oil.
**oleoresins** — extracted, concentrated, and standardized essential oils and nonvolatile components of spices; usually in the form of a paste or solid.
**oleostearin** — or, "stearin." The solid fraction of edible beef fat, rendered tallow, and the like.
**olie** — *Netherlands Denmark* oil.
**oliebol** — *Netherlands* fritter containing raisins.
**olio** — *Italy* oil.
**olive oil** — oil pressed from the fruit of the olive tree. It is high in unsaturated fat and has a characteristic flavor that is much liked by some consumers and disliked by others.
**olja** — *Sweden* oil.
**öljy** — *Finland* oil.
**olut** — *Finland* beer.
**omochi** — *Japan* rice cake.
**onion rolls** — bread rolls with chopped onions sprinkled on top before baking or, less often, distributed throughout the unbaked dough.
**onions** — the plant *Allium cepa*, and its bulb. Many varieties are grown commercially and used as an edible vegetable and condiment. Available as ingredients in fresh, frozen, and dried forms.
**ontbijt** — *Netherlands* breakfast pastry, honey cake.
**opacity** — degree of resistance to transmittance of light.
**operations research** — a scientific method utilizing statistical analysis for studying and improving the operations of men and machines.
**optical sensors** — electrical or electronic devices that cause a change in electrical output when there is some modification in the visible light impinging upon them.
**orace** — *Serbia/Croatia* walnuts.
**organic acids** — organic compounds bearing one or more unreacted carboxy groups; includes such common food acids as acetic, citric, and tartaric.
**organoleptic** — external stimuli detectable by the senses, e.g., flavor, odor, color, etc.
**orifice meter** — a thin plate with a hole of accurate size bored in it;, when combined with gauges upstream and downstream to measure fluid pressure, can provide a determination of rate of flow in a pipe.
**orus** — *Egypt* a yeast-leavened breakfast pastry consisting of a rich but not sweet dough folded over a chopped date filling which has been fermented overnight, then baked.
**Osborne's method** — a scheme for classifying wheat proteins based on their dispersibility in various solvents.
**osmophilic** — describes microorganisms that can function in systems having a low water activity (jellies, high solids syrups, etc.).
**ost** — *Denmark* cheese.

**ostkaka** — *Sweden* a type of cake containing a large proportion of cheese curd.

**oststänger** — *Sweden* cheese straws (thin bread sticks flavored with cheese.)

**othello** — confection made from a highly aerated mixture of egg whites, egg yolks, and sweetener that is formed into small mounds and baked on paper. Sometimes coated with chocolate after cooling.

**othellokage** — *Denmark* layer cake filled with custard, topped with chocolate sauce and whipped cream.

**ounakook** — *Estonia* apple cake.

**ounamungad** — *Estonia* similar to apple dumplings, but the apples are wrapped in a lean yeast-leavened dough.

**ounce** — a measurement of weight; in the absence of a modifier, this means an avoirdupois ounce equivalent to 28.3495 grams. Troy ounces and apothecary ounces are equal to 31.0135 grams.

**ounce, fluid** — a measurement of volume, equal in the US, to 29.6 cc, but in the UK equal to 28.4 cc.

**oven** — a heated chamber or partially inclosed space used for cooking foodstuffs with heat transferred by means other than by contacting the product with a liquid heat transfer medium. There are many oven shapes, sizes, and heating and conveying means.

**oven finished** — describes cakes, and some other baked goods, prepared from batters that have had fillings or icings deposited on them before they are baked; a variation is the case where batter has been deposited on top of a kind of filling before baking.

**oven hook** — a hook with a long stem attached to a wooden handle; used for reaching into a hot oven to pull out pans.

**oven loaders** — machines that transfer pans or straps from an incoming conveyor to an oven conveyor, sometimes arranging the pans in various patterns to achieve maximum utilization of oven space.

**oven-puffed** — applied to grains (principally rice) that have been moderately expanded by rapidly heating moist partly gelatinized kernels at ambient pressure, usually in an oven. Contrasted to gun puffing and extrusion puffing, which involve pressurization and pressure-release steps applied to gelatinized grains or pellets.

**oven sheet** — a recording form on which the oven operator enters important data regarding conditions affecting the baking product.

**oven spring** — the expansion of the loaf or other dough piece that occurs during baking.

**over-and-under scale** — a weighing device on which the indicator shows the weight as being either over or under a pre-set figure. Useful for rapid on-line determination of weight compliance.

**overhead proofer** — equipment for holding and transporting dough pieces (usually through an inclosed chamber) during the intermediate proofing period; so-called because it is usually installed near the ceiling to make the best use of bakery space.

**overproof** — deterioration of dough resulting from excessive proof time and/or temperature.

**ovnbagt** — *Denmark* baked.

**ovos** — *Portugal* eggs.

**oxidant** — an oxidizing compound, i.e., a substance that causes oxygen to be added to another compound or deprives a compound of hydrogen. In baking, certain oxidants such as bromates and iodates are particularly useful in modifying dough properties.

**oxidation** — a chemical reaction involving the addition or combination of oxygen with another material or, more generally, an increase in the number of positive charges on an atom or a reduction in the number of negative charges. Oxidation in food products containing fat can result in the development of rancidity with accompanying objectionable flavors and odors.

**oxidizer** — see "oxidant."

**oxidative rancidity** — the unpleasant smell, often accompanied by a bad taste, that occurs when a fat has been oxidized, i.e., when the fatty acids have acquired oxygen atoms at their double bonds.

**oxygen scavenger** — a material that can absorb or otherwise remove traces of oxygen from a hermetically-sealed container.

**ozone** — an allotropic form of oxygen, having the molecular formula $O_3$ and possessing the ability to destroy microorganims at low concentrations; particularly useful in sterilizing potable water.

## -P-

**packed tower aeration** — or, "air stripping." A water purification treatment for reducing the content of volatile organic compounds to acceptably low levels.

**paczki** — *Poland* buns or rolls of fried yeast-leavened dough, similar to doughnuts but without the hole; usually filled with, e.g., jam after frying.

**paddle beater** — a mixer agitator of the batter beater type.

**paella** — *Spain* cooked rice (flavored, as with saffron) mixed with assorted seafood and, sometimes, meats.

**pähkinä** — *Finland* nut.

**pähklitort vahukoorega** — *Estonia* walnut torte

**pain** — *France* bread; a loaf.

**pain complet** — *France* whole wheat bread.

**pain de campagne** — *France* country-style bread, usually a round hearth baked loaf weighing several pounds.

**pain de gruau** — *France* white bread made with the very best available flour.

**paindemaigne** — *Old English* original name for manchet.

**pain de ménage** — *France* family loaf; regular bread.

**pain de mie** — *France* loaf baked in an inclosed pan (a la Pullman) to produce a loaf of closely defined dimensions and cross-section.

**pain ordinaire** — *France* common or standard bread.

**pain perdu** — a slice of white bread covered with a rich batter and fried.

**paj** — *Sweden* pie, tart.

**palacinke** — *Serbia/Croatia* pancakes.

**palacsinták** — *Hungary* pancakes, usually served with a filling.

**palacsinta tészta** — *Hungary* pancake batter.

**palatschinken** — *Germany* pancake usually filled with jam or cheese, and often served with a hot chocolate and nut topping.

**palenta** — *Serbia/Croatia* cornmeal mush.

**palette knife** — a spatula, generally thin and flexible with a rounded end, used for spreading icings and the like on cakes and other baked products.

**pallet** — a platform generally made of a cheap grade of lumber, having two layers of boards between which the forks of a lift truck can fit. They are used as bases on which containers of ingredients and products can be stacked for storage or transfer. Some are made of plastic.

**palmiers** — a sweet pastry baked from a cross-section of a two-lobed puff pastry cylinder shaped so as to roughly resemble a palm frond.

**palm nut oil** — also, "palm kernel oil." Oil obtained from the kernel or seed of the fruit of the oil palm.

**palm oil** — oil pressed from the fleshy part of the fruit of the oil palm.

**palm sugar** — a thick, crumbly, strong-flavored sugar found in various brown hues; it is made by evaporating the sap from palmyra palms or sugar palms, and is not a common item in international commerce.

**pan** — *Spain* bread.

**panada** — *Spain* (1) A paste based on bread or flour, used as a binding ingredient in croquettes or the like. (2) A soup made of bread, broth, and butter.

**panaderia** — *Spain* bakery.

**panadero** — *Spain* baker.

**pan blanco** — *Mexico* white bread.

**pan bread** — loaves baked in pans or tins, as contrasted to hearth bread.

**pancakes** — relatively thin, nearly always round cakes, of widely varying sizes, baked on a griddle from batters of thin to moderately thick consistency. The batter is usually chemically leavened, but there are popular versions based on yeast-leavened, sour dough, or vapor-leavened (egg foam) mixtures. Can be served with butter and syrup (as they usually are in the U.S.), wrapped around sweet or savory fillings, or cut up into strips for use in soups and the like.

**pan centeno** — *Spain* brown bread; rye bread.

**pan crust** — the part of the crust of the loaf that has come in direct contact with the inside surface of the pan during baking.

**pan de caja** — *Mexico* loaf of bread which has been baked in a rectangular pan.

**pan de centeno moreno** — *Mexico* pumpernickel bread.

**pan de Genova** — *Italy* almond cake.

**pan de huevo** — *Mexico* yeast-leavened buns containing eggs, generally less rich than pan dulce and often flavored with cinnamon. Shaped as shells (conchas), for example.

**pandekage** — *Denmark* pancake.

**pan de pasas** — *Mexico* raisin bread.

**pan de Spagna** — *Italy* sponge cake.

**pandowdy** — a dessert dish consisting of sweetened and spiced apple slices that have been covered with deposits of rich biscuit dough or streusel, then baked.

**pan dressing** — a thick mixture (consisting typically of sugar, butter, nuts, etc.) that is spread on the bottom and sides of a baking pan, then the batter or dough deposited in the pan and baked.

**pan dulce** — *Mexico* sweet dough buns in various flavors and shapes; usually they do not have icings or other decoration but the dough may be colored.

**pane** — *Italy* bread.

**pané** — *France* breaded, rolled in bread crumbs.

**panecillo** — *Mexico* bread roll.

**pane di segale** — *Italy* rye bread.
**panerad** — *Sweden* breaded.
**paneret** — *Denmark* breaded.
**pane scuro** — *Italy* dark/black bread.
**panetone** — *Mexico* a simpler, less rich, version of the Italian panettone; a pastry containing candied fruits.
**panettone** — *Italy* a rich yeast bread of traditional (tall, cylindrical) shape containing fruits; an Italian Easter specialty. The original method of processing is very elaborate and time-consuming.
**pan fino** — *Spain* generic term for Mexican-style pastries made of a fairly lean dough and decorated with a baked-on paste made of flour, sugar, water, shortening, colors, and flavors.
**pan flare** — the extent to which the side of a rectangular bread pan is angled outward from the bottom.
**panforte di Siena** — *Italy* flat round slab made mostly of spiced crystallized fruit.
**pan francés** — *Mexico* hearth-baked loaf bread, made from lean dough, usually having a split top.
**pan glaze** — a semi-permanent coating (usually a silicone) applied to the inner surface of loaf pans to reduce sticking of baked products.
**pangrattato** — *Italy* breadcrumbs.
**pan grease** — generally, compounds of food oils and/or mineral oils with various additives; used for coating the insides of pans so the baked products can be easily removed.
**pan greaser** — a machine that automatically applies (by wiping or spraying) a temporary coating of oil or grease to inner surfaces of baking pans.
**pan handling systems** — an assemblage of machines that stack pans or straps that arrive from the depanning station, move stacks to storage, remove stacks from storage, deliver pans to the depositer as required, coordinating the movement to demand from the manufacturing line.
**paniert** — *Germany* breaded.
**panino** — *Italy* bread roll.
**panko** — *Japan* dried, toasted bread crumbs, of relatively large particle size, used for coating fried foods.
**pan liners** — shaped forms of paper or parchment, sometimes treated with non-stick additives such as silicones, to be inserted in baking pans for the purposes of facilitating product removal and assisting in maintaining product integrity during transfer.
**panna** — *Italy* cream.
**panna montata** — *Italy* whipped cream.
**pannekoek** — *Netherlands* pancake.
**panning** — the process of placing dough pieces in baking pans, whether done manually or by machines.

**pannkaka** — *Sweden* pancake.
**pannkoogid** — *Estonia* pancakes.
**panocha** — *Mexico* a circular loaf of bread, perhaps 10 or 12 inches in diameter and about 1 or 2 inches thick, usually made from a lean chemically leavened formula and cooked in a skillet over an open fire; traditionally made on the trail by cowboys, deer hunters, and other campers.
**panoche** — also, "panocha." *Mexico* (1) A kind of raw sugar. (2) A confection generally made from partially refined sugar, cream, and nuts, often in the shape of a medium-sized circular cookie.
**pan proofers** — the complex of equipment that transfers baking pans and their contents from the depositing station, holds them under controlled atmospheric conditions and temperature for a predetermined time, then delivers them to conveyors leading to the oven. For manual operations, the chamber that holds raw dough in pans until final proofing is completed.
**panque nuez** — *Mexico* cake-shaped sweet bread topped with pecans.
**panqueque** — *Mexico* pancake.
**pan rack** — a stationary or movable structure of open shelves on which baking sheets or pans may be placed.
**pans** — variously shaped metal containers for baking or cooking.
**pan stacker** — machine for placing pans (or straps) that are temporarily out of use in stacks for efficient use of storage space.
**pan tostado** — *Spain* sweet toasted white bread.
**pan tostato** — *Italy* toasted bread slices, often spiced.
**pan toasted** — as applied to flaked oat groats packaged for retail sale as a to-be-cooked breakfast cereal, describes a heat treatment in open pans or conveyors, during which the particles undergo a very slight browning.
**pan washer** — a specialized automatic washing device for cleaning baking pans and similar utensils.
**panzarotti** — *Italy* large dough pockets filled with, e.g., pork, eggs, cheese, and tomatoes; cooked by frying or baking.
**pao** — *China* steamed Chinese bun. *Portugal* bread.
**pao de centeio** — *Portugal* rye bread
**pao de ló** — *Portugal* coffee cake.
**paozinho** — *Portugal* bread roll
**papa** — *Mexico* potato.
**papain** — a protein-digesting enzyme obtained from papaya fruit; it is much used in meat tenderizers and has been used as a gluten softener, but was largely displaced from the latter application by fungal enzymes.
**paperboard** — a relatively thick form of solid (as opposed to corrugated) paper-like material, often coated, printed, and laminated, that is the basic material for most bakery product boxes.
**pappadams** — also, many variant spellings. *India* Round pieces of flatbread fried in shallow oil until crisp; thin but of fairly large diameter,

usually made from lentils, but sometimes from rice or potato flour. May be flavored with cumin and/or pepper. Served as snacks.

**pappardelle** — *Italy* long, broad noodles.

**paprika** — pulverized dried pods of *Capsicum annum*, a sweet red (bell) pepper. Valued for the brilliant red color it can contribute to foods, it is also used for its flavor.

**parabens** — para-hydroxybenzoic acid esterified with methyl, ethyl, propyl, or butyl alcohol; inhibitors of microbiological spoilage.

**paratha** — also, "parantha." *India* An unleavened dough ball made with whole wheat meal is coated with fat (ghee), then flattened and fried on a hot iron plate. Often shaped as thin triangles.

**parboiling** — (1) In general culinary operations, the boiling for a short time of a food piece (such as a raw chicken leg) as a preliminary to further cooking at some later time. (2) In rice processing, a method for facilitating the release of the hull, transferring some nutrients from the hull to the kernel, and obtaining other changes, that involves the essential steps of wetting the rough rice, heating it, then drying it.

**parfait** — ice cream and syrup (or whipped cream) placed in alternate layers in a serving glass. Also used fancifully to describe striped, filled hard candies.

**Parker House rolls** — bread rolls of about 2 oz weight formed by folding 1/3 to 1/2 of a thin rectangular dough strip over the rest of strip before baking.

**parkin** — *UK* a cake made principally of oatmeal, containing also ginger and molasses.

**parmesan** — a hard Italian cheese with pungent aroma; most of it is used in grated form. Often dusted on cheese sticks, garlic toast, spaghetti, etc.

**pärmi** — *Estonia* yeast.

**pasas** — *Spain* raisins.

**paskalya cöregi** — *Turkey* a lightly sweetened, yeast-leavened cake or loaf, an Easter specialty.

**påskefestbrod** — *Denmark* yeast-leavened, slightly sweet dough containing raisins, citron, etc., and baked in loaf form for Easter meals.

**passa de uva** — *Portugal* also, just "passa." Raisin.

**passatelli** — *Italy* dough formed from a mixture of egg, parmesan cheese, breadcrumbs, and nutmeg; used in some of the same applications as semolina pasta.

**pasta** — a generic name for products such as macaroni, spaghetti, and noodles; also called alimentary pastes, macaroni products, noodles, etc.; all generally made of durum semolina dough that has been extruded or sheeted and cut; found in hundreds if not thousands of different shapes. *Mexico* Dough. *Greece* Pastry, plain cake.

**pasta asciutta** — *Italy* any pasta not eaten in a bouillon.

**pasta frolla** — *Italy* a sweet pastry very much like a savarin.
**pastas** — *Spain* macaroni products, alimentary pastes.
**paste** — term generally applied to a "dead" (unleavened, non-elastic) coherent mixture; almond paste is an example.
**pastei** — *Netherlands* pie, pasty.
**pasteija** — *Finland* pastry, pie.
**pastej** — *Sweden* pie or tart.
**pastel** — *Spain* cake, such as a layer cake or a pastry such as pie. *Portugal* most often, a pie but occasionally a croquette or other shape, not always made of dough.
**pastel de Belem** — *Portugal* custard pie.
**pastel de cereza** — *Spain* cherry pie.
**pastel de chocolate** — *Mexico* chocolate cake, usually layer cake.
**pastel de compleaños** — *Spain* birthday cake.
**pastel de folhado** — *Portugal* a pie, tart, etc. made with puff pastry dough.
**pastel de manzana** — *Spain* apple pie.
**pastel de queso** — *Mexico* also, "quesadilla." Cheese cake.
**pastel de Santa Clara** — *Portugal* small tart with almond paste filling.
**pasteles** — *Mexico* pastry.
**pastelillo** — *Spain* small tart.
**pastete** — *Germany* pastry, pie.
**pasteurization** — relatively mild heat treatment sufficient to kill vegetative forms of the predominant spoilage organisms in foodstuffs; not synonymous with sterilization.
**pasticcino** — *Italy* tart, cake, small pastry.
**pasticcio** — *Italy* pie. Also, a type of pasta similar to lasagne.
**pastie** — also, "pasty." A savory pie of the Welsh type, generally made in single-serving size. They usually have a non-flaky crust and, often, a beef-onion-potato filling; in shape like a fried pie, but baked.
**pastillage** — *France* gum paste used in making candies, decorations, etc.
**pastina** — *Italy* any of the small pasta pieces used principally as soup ingredients.
**pastisetas** — *Mexico* octagonal-shaped butter cookies.
**pastitsio** — *Greece* similar to baked lasagna.
**pastry** — an inexact term that is now generally regarded as applying to almost any dessert-type baked food.
**pastry bag** — a conical canvas (or plastic) bag that narrows to a small open point at one end and has a large opening at the other end; it is filled with a decorating paste (or deposit cookie dough), the larger open end is folded over and held closed, and hand pressure is used to force out the contents in a narrow strand from the pointed end; usually, a metal or plastic orifice is inserted in the pointed end to give some decorative treatment to the extruded material.

**pasty** — *UK* a baked pie of the turnover type, usually filled with chopped cooked seasoned meat.
**patakukko** — *Finland* pie with rye flour crust filled with whitefish and bacon.
**patata** — *Spain* potato; *Mexico* sweet potato.
**patates tsips** — *Greece* potato chips.
**pâte** — *France* dough or paste.
**pâté** — seasoned meat or fish ground into a paste consistency and cooked, often in an earthenware dish called a terrine.
**pâte à chou** — *France* cream puff paste.
**pâte à dumpling** — *France* the dough for dumplings, often consisting only of flour, water, and seasoning (unleavened).
**pâte à pâté** — *France* any of the various kinds of unleavened dough sheets used to cover, for example, meat pies. Often a simple lard, flour, salt, and water mixture.
**pâte bâtarde** — *France* standard formula dough for making ordinary white bread.
**pâte brisée** — *France* dough similar in composition to pie crust, not sweet but sometimes contains spices.
**pâte feuilletée** — *France* puff pastry dough.
**patent flour** — product made from the finer and whiter flour streams; comes mainly from ground purified middlings. Has lower bran content and higher endosperm content than straight flour, thus lower ash and lower fiber content.
**pâtes** — *France* macaroni products, alimentary pastes.
**pâte sablée** — *France* rich sweetened shortcrust pastry, as used for certain sweet tart crusts.
**pâte sucrée** — *France* dough similar in composition to pie crust, but sweeter due to the addition of 10% to 15% powdered sugar.
**patispanj** — *Serbia/Croatia* sponge cake.
**patisserie** — pastry.
**patonki** — *Finland* French bread.
**patty shell** — a pre-baked casing for savory preparations such as creamed chicken; in shape, typically a short vertical cylinder with an open circular cavity in the top center; often made of puff pastry dough; similar to a vol au vent.
**pavlova** — *Australia* a base of hard meringue, slightly baked then filled or covered with a fruit preparation (many variations) and topped with whipped cream. Has been called the Australian national dessert.
**pay** — *Mexico* pie.
**peak** — (1) Stage of maximum goodness. (2) In preparing meringues and some other whipped goods, it is the stage at which the mixture will form a persistent point when the beater is pulled vertically from the beaten

mixture. A "dry peak" is short and stiff and has a rather dull appearance, a "wet peak" is long, soft, bending, and has a rather glossy appearance.

**peaked** — a cake or other baked product that rises up in the center to an excess degree, giving a somewhat pointed appearance.

**peanut** — the ground nut, goober peas. The plant is actually a legume, but its seed is similar in many ways, including edible usage, to tree nuts. Common varieties are Virginia, Spanish, and Runner. Shelled peanuts are roasted before use, except in one or two uncommon applications.

**peanut brittle** — a confection made by caramelizing a molten sugar and butter mass containing peanuts, then mixing in a small amount of sodium bicarbonate before sheeting out to a thickness of about one-fourth inch. The sheet is cooled until it becomes brittle, then broken into pieces.

**peanut butter** — a paste produced by milling roasted (and usually blanched) peanuts to a fine particle size. Hydrogenated vegetable oil and emulsifiers are usually added to prevent oil separation and modify the texture. Salt is always added, except in dietetic products.

**peanut oil** — oil pressed from unroasted shelled peanuts. It is, under normal market conditions, too expensive to compete with soybean or cottonseed oils for routine bakery use so it is normally restricted to specialty and gourmet applications.

**pearling** — an abrading process applied to many grains, but primarily barley, to remove the outer coatings of the grain and convert the kernel into a more or less spherical form; the "pearls" are used in soups, etc., or, less often, as an intermediate in a flour/meal milling process.

**pearling index** — a test involving a pearling operation applied to wheat and other grains to allow an estimation of the hardness of the kernels.

**pearl millet** — a type of millet (*Pennisetum glaucum*) used as a food in parts of Africa and Asia, and grown as a feed crop elsewhere.

**pecans** — seeds of the tree *Carya illinoensis*. The nutmeats have a slight resemblance to small walnut meats, but their length to width ratio is greater. The flavor is mild and pleasant. A premium nut, primarily restricted to North American use. Does not need to be roasted before use in bakery products. Obligatory in pecan pies, of course.

**pectin** — a hydrocolloid commercially obtained from citrus peel and apples. Consists mainly of partial methyl esters of polygalacturonic acid. Gels sugar solutions under certain conditions, leading to formation of jellies, jams, etc. A "dietary fiber" substance, although it doesn't form fibers.

**peel** — wooden or metal paddle with a long handle and a wide flat blade used to place loaves (or pans) in an oven and take them out.

**peel loader** — machine for automatically loading baking sheets into traveling-hearth ovens.

**peel ovens** — large stationary ovens, so called because the wide and deep baking hearths have to be loaded and unloaded with peels.

**Pekar test** — see "slick test."

**pekmez** — *Serbia/Croatia* jam.

**peksimet** — *Turkey* similar to bazlama (q.v.) except baked in a peel oven.

**pelican** — a spout sampler for obtaining a representative sample from a falling stream of grain.

**pelle** — *France* peel.

**pelleting** — in the preparation of feedstuffs, the operation of compressing the mixture into pieces of more or less uniform size, for the primary purpose of preventing separation of the microingredients.

**pelmeny** — *Russia* a stuffed noodle dumpling, virtually the same as "pierogi" in composition and use.

**pendulum-balanced scales** — weighing devices that compare an unknown mass with the force exerted by a known weight mounted on the end of a rigid rod that rotates about a pivot.

**pentosan** — a polymer of pentoses, the latter being five-carbon sugars; the pentose analog of starch.

**peperkoek** — *Netherlands* gingerbread.

**pepparkakar** — also, "pepparkakor," *Sweden* gingerbread, spiced cookies, usually soda-leavened.

**pepper** — (1) Common black pepper, the dried seed of a climbing vine cultivated mostly in Brazil and India. (2) White pepper is based on the same raw material, but the seeds are decorticated, so that only the white inner portion is ground. (3) Red pepper spices, of which cayenne pepper is a common example, consist of the dried fruits of certain plants belonging to the genus *Capsicum*. They are added mostly for the hot, stinging sensation or taste they contribute to foods, although they are also useful as colors. (4) Bell peppers are used mostly as the fresh fruit, in salads and the like; also produced by the genus *Capsicum*, but many varieties have little or no heat.

**peppermint** — possibly the most common variety of mint, q.v.

**peppernuts** — pfeffernusse.

**PER** — protein efficiency ratio, based on the amount of a specific protein, such as casein, that is required to provide the minimum amount of essential amino acids.

**pera** — *Spain* pear.

**pericarp** — botanically, the ripened and variously modified walls of the ovary. In the cereal grains, it is generally observed as a thin and foliaceous or membraneous layer covering the kernel. Separated in wheat milling as the bran fraction.

**perlgraupe** — *Germany* pearled barley.

**permeable** — describes a film or sheet of a material through which molecules or ions can pass, usually selectively.

**peroxide value (PV)** — oxygen can react with a fatty acid chain to form peroxides or hydroperoxides. Peroxide value is a measure of the amount of these materials that have been formed in a sample. It is expressed as milliequivalents of peroxide-oxygen combined per kilogram of fat (meq/Kg).

**perunalastut** — *Finland* potato chips.

**pet foods** — foods designed to meet the complete dietary needs of some category of pets, such as dogs, cats, canaries, ornamental fish, or gerbils.

**petit bouchée** — a miniature patty shell of puff pastry used with savory fillings to make a single-bite hors d'oeuvre.

**petites galettes salées** — *France* small salted crackers.

**petit fours** — according to current US usage, small chemically leavened cakes, usually square, covered with colorful (and usually decorated) icings. Thin cake layers are often alternated with jam or other fillings to make rather elaborate assemblages. They are frequently based on lean pound cake formulas or firm layer cake slices. The term is never applied to multiple serving pieces.

**petit pain** — *France* bread roll or small loaf.

**pétrissage** — *France* the kneading process.

**pfannkuchen** — *Germany* pancake.

**pfeffernusse** — also, "peppernuts." A Christmas cookie popular in Germany. In its usual versions, appears as a small ball-shaped cookie coated with powdered sugar. Texture crisp. Has a sharp spicy taste. Formulas usually include a small amount of black or white pepper.

**pH** — measure of the acidity or alkalinity of an aqueous system, pH 7 being neutral while lower values denote acidic conditions and higher values imply alkalinity; in actuality, the negative logarithm to the base ten of the hydrogen ion concentration of the solution.

**Philadelphia butter cake** — a coffee cake made by baking sweet dough in a pan lined with a dressing consisting principally of sugar, butter, eggs, and NFDM.

**phosphatase** — an enzyme acting primarily to split esters of orthophosphoric acid.

**phosphated flour** — wheat flour to which monocalcium phosphate has been added to serve as the acidic component of a leavening system, i.e., only soda has to be added by the user to cause the dough or batter to rise.

**phospholipids** — organic compounds with a lipid moiety attached to a phosphoric group, lecithin being an example. Many phospholipids have some emulsifying action.

**phytase** — enzyme that breaks down phytin into smaller molecules.

**phytin** — phosphorus compound found in wheat bran (and elsewhere) that is one form of dietary fiber; it binds some metallic ions.

**phytoglycogen** — a polysaccharide consisting of glucose residues linked by alpha-D-(1,4) bonds, with branching at alpha-D-(1,6). It may constitute as much as 25% of the dry weight of a sweet-corn kernel.

**piccante** — *Italy* highly seasoned.

**picker-shellers** — machines for harvesting popcorn and the like; they remove the ear from the stalk, peel off the husk, and strip the kernels from the cob.

**pidesi** — *Turkey* flatbreads in general.

**pie** — pastry shell (usually round, with a raised sloping rim) containing a filling of sweetened fruit, pastry cream, custard, etc. which is often topped with a layer (or strips) of pastry or with whipped cream, meringue, streusel, etc. Usually, the crust is unleavened. Thousands of variations are known, including meat pies, fried pies, etc.

**piebald kernels** — grain, and particularly wheat kernels, that show patchy separation of the endosperm into vitreous and floury sections, a pattern also visible on the surface.

**pie pin** — a thin, relatively long rolling pin used for sheeting pie doughs,

**pie press** — a machine for automatically forming pie crusts out of lumps of dough by forcing the dough into a mold.

**pie rimmer** — a device for trimming the excess dough from the edge of a crust, sometimes provided with attachments to form decorative (crimped) rims.

**pierniczki** — *Poland* cookies or small pastries.

**piernik** — *Poland* gingerbread, also cake, etc.

**pierogi** — *Poland* similar to ravioli; a dough of varying composition folded or otherwise placed around a sweet or savory filling such as mashed potatoes, cheese, sauerkraut, or plums.

**pigment** — a substance that imparts a color (including black and white); both natural and artificial pigments are used in foods.

**pignolis** — small white nuts taken from a tree of the pine family.

**piima** — *Estonia* milk.

**piirakka** — *Finland* pie.

**piiras** — *Finland* small pie or pasty.

**pikelet** — *UK* a griddle-baked bread product very similar to English muffins and crumpets, the differences not being clearcut or universally agreed upon.

**pikkuleipä** — *Finland* cookie.

**pilaf** — several variant spellings are current. A casserole-type preparation of cooked rice (sometimes cracked wheat or bulgur) combined with minor amounts of various other ingredients such as meat or vegetables, seasonings, spices, and condiments.

**pilâvlar** — *Turkey* pilafs.

**pils** — *Netherlands* beer (generic term).

**piña** — *Spain* pineapple.

**piña colada** — *Cuba* although this term was originally applied to a rum drink, the combination of coconut and pineapple flavors (sometimes with rum) has been used extensively in fillings, frostings, and icings for cakes.

**pinch** — an inexact ingredient measurement, considered to be about one-eighth teaspoon.
**pinda** — *Netherlands* peanut.
**pindakaas** — *Netherlands* peanut butter.
**pineapple** — fruit of a tropical plant, grown in large amounts in Hawaii and Mexico. For the baker, canned pineapple (sliced into discs or wedges, or crushed) is the ingredient most often employed, but candied pineapple is required for fruit cake.
**pinipig** — *Philippines* toasted and flattened glutinous rice that is used in cakes and desserts.
**pinoccate** — *Italy* cake containing pine kernels and almonds.
**pint** — in the U.S.A., 16 fluid ounces; in the UK and Canada, an Imperial pint is 20 fl oz.
**piparkakku** — *Finland* gingerbread.
**piparkoogitort** — *Estonia* gingerbread tort.
**pipe** — (v) to extrude cream puff paste, filling, icing, or piping jelly out of a either a piping bag or a small hand pump.
**piperine** — an alkaloid found in black pepper, and elsewhere, that causes a hot sensation in the mouth.
**piping bag** — a utensil very similar (if not identical) in shape and method of use to a pastry bag, but generally smaller.
**piping jelly** — soft gels, usually colored but seldom flavored except with sweeteners, that are used to decorate cakes, etc.; applied in thin strips with a piping device or to larger areas with a spatula.
**pippuri** — *Finland* pepper.
**pirog** — *Russia* pie, flan, tort. *Sweden* pasty filled with caviar, cheese, fish, or vegetables.
**pirogen** — Yiddish equivalent of "pierogi."
**pirozhki** — *Russia* baked pies of the turnover or pasty type made with leavened dough and filled with fish, onion, and/or mushrooms.
**pirozhnoye** — *Russia* flaky pastry with rich cream filling, often like a napoleon.
**pirukatäited** — *Estonia* pastry filling.
**pirurutung** — *Philippines* dark, purplish non-glutinous rice.
**piskota** — *Serbia/Croatia* sponge cake.
**piskótatészta** — *Hungary* sponge cake.
**pissaladière** — *France* a type of flan said to be "first cousin to an Italian pizza;" a specialty of Nice.
**pistachios** — small to medium sized nuts, the meats being variegated greenish and light brown with a distinctive flavor. The shells are normally ivory to white in color, but are often dyed red. Formerly, most pistachios were obtained from Iran, but superior quality nuts are now available from U.S. sources. Less subject than other nuts to rancidity.

**pistolet** — *France* split roll, something like a club roll.

**pita** — *Serbia/Croatia* flaky pastry in a roll with a filling, such as cheese.

**pita bread** — a thin, circular yeast-leavened bread that, due to the method of baking, has the top and bottom crusts substantially completely separated except at the edges. Used to form pockets that are filled with savory preparations.

**pithiviers** — *France* a round, flat cake having layers of puff pastry alternating with almond cream or other sweet fillings.

**pitot tube** — a device for measuring fluid flow, consisting essentially of a cylindrical pipe having a short, right-angled bend, placed vertically with its mouth normal to the direction of flow in a moving stream of fluid.

**pitta** — *Greece* a round flatbread made from a yeast-leavened lean dough.

**pivo** — *Russia* beer.

**pizza** — a flat dough piece on which additives such as cheese, sausage slices, and tomato sauce can be placed; usually round, thin, and baked, but thousands of variations have appeared.

**pizza press** — a device for compressing a weighed amount of pizza dough into a circular shape, sometimes with a rim.

**pizza rolls** — small savory pies of the egg roll type, usually flavored with a combination of spices reminiscent of pizza flavor.

**pizzetta** — *Italy* small pizza.

**placek serowy** — *Poland* cheese cake.

**placuszki z macy** — *Poland* matzo fritters; matzos soaked in boiling water and combined with beaten egg whites, then with beaten egg yolks; formed into pieces and fried.

**plait** — a way of forming fancy breads and sweet goods; consists of intertwining strips or strings (usually three) of dough.

**planetary mixers** — also, "planetary-action vertical mixers." Vertical mixers in which the agitator revolves around its own axis while the axis itself describes a circle about midway between the center and the side of the bowl.

**plansifter** — a machine in which a number of oscillating rectangular sieves, aligned horizontally or at a slight angle and arranged one above the other, are used to separate the components of a mill stream.

**plant exudate gums** — natural gums based on the sap extruded from certain trees and shrubs which is collected, purified, and dried to form food ingredients and the like. Many ancient examples, such as gum arabic.

**plaque** — *France* a sheet of metal or other material on which bread or the like is baked.

**plasticity** — the consistency (texture or "feel") of a solid or semi-solid shortening.

**plasticizing** — a method of treating a shortening or other fatty-type of ingredient so as to develop a crystal structure that has improved processing response and organoleptic characteristics, usually done by mechanically working the plastic material within carefully controlled temperature ranges.

**plastic range** — the range of temperatures within which a fatty material appears to be a solid mass but can be readily deformed without breaking; the upper and lower temperatures between which a shortening has a suitable consistency for use in creaming and the like.

**plastic resins** — or, "resins." The more-or-less pure chemical preparations that are used as the raw material for preparing plastic objects; polyethylene and polystyrene are examples of resins.

**plátanos** — *Spain* bananas, especially those of the large green type cooked as a vegetable.

**platform scales** — devices that measure the weight of objects placed on a large, flat, often rectangular, surface that rests on sensing elements.

**plath** — a Welsh type of flatbread.

**plating** — (1) Applying a very thin layer of dissimilar metal to a metal object. (2) Depositing a thin layer of flavoring or coloring material on to fairly large crystals, as of granulated sugar; usually accomplished by gentle mixing of the crystals with a concentrated solution of the additive, which is then allowed to dry. (3) Preparing a culture of microorganisms on, say, agar plates.

**plättar** — *Sweden* small thin pancakes.

**plätzchen** — *Germany* cookie.

**playback robots** — mechanical task-performers with electronic controls that must be programmed by manually taking the arm and gripper (or other elements) through the task step-by-step.

**pliable** — describes a dough that has a good consistency for processing, generally understood to mean soft but extensible.

**plighuri** — *Greece* cracked wheat noodles.

**plogues** — *Canada* little crepes made of white buckwheat flour, to be folded with butter and eaten from the hand.

**plows** — stationary curved metal appliances, fixed above a conveyor line on which long or continuous dough strips are being carried; for the purpose of turning over part or all of the dough strip or diverting it.

**plum duff** — a plain flour pudding containing raisins, cooked by wrapping it in a cloth bag and boiling it.

**plum pudding** — a blend of raisins, currants, citrus peels, milk, suet, and spices, with just enough flour to bind these ingredients together; cooked by boiling or steaming, the pudding being held in a cloth. Many variations have been described, including baked types.

**pneumatic conveyors** — systems that suck or blow powdered or granular materials through pipes for the purpose of transferring them from, e.g., a bin to a dispensing station.

**pneumatic equipment** — devices relying on controlled pressurized air for their motive power.

**pocket** — in a divider, the cavity into which the dough is pressed for volumetric measurement.

**pod corn** — a variety of maize in which the individual kernels are inclosed in husks. Not a commercially important type.

**poemat** — *Poland* a cross between pudding and unbaked cheese cake.

**poffertje** — *Netherlands* fritter to be served with sugar and butter.

**pogaca** — *Serbia/Croatia* round flat bread, in some cases unleavened.

**pogacice od cvaraka** — *Serbia/Croatia* a kind of cracker containing some crackling.

**pogácsa** — *Hungary* biscuits in general; yeast-leavened or unleavened.

**pogne** — *France* rich bread, similar to brioche; sometimes filled with candied fruit.

**pohovano** — *Serbia/Croatia* breaded.

**pointage** — *France* bulk fermentation.

**polenta** — *Italy* (1) a thick porridge of cornmeal, semolina, barley meal, or ground chestnuts mixed with water and minor flavorings such as butter, salt, and grated Parmesan cheese. (2) The mush as previously described made into squares or small balls and browned in hot fat.

**Polish wheat** — a variety of *Triticum turgidum* having large loose spikes with conspicuous papery glumes and very hard large, long, yellowish-white kernels resembling those of durum. Formerly, *Triticum polonicum*.

**polulebli hleb** — *Serbia/Croatia* bread made from unbleached flour.

**polvore crescente** — *Italy* baking powder.

**polvorones** — *Mexico* a type of pan dulce, usually in the form of small, circular, moderately thin buns or thick cookies. May be flavored, as with chocolate. Leavened with bakers' ammonia or baking powder, but dense.

**polydextrose** — a nonsweet polymer of glucose that is primarily useful as a bulking agent. Since the claimed energy yield is one calorie per gram, it could be a useful reduced-calorie substitute for starch.

**polyester** — usually refers to polyethylene terephthalate, a plastic resin much used for packaging applications requiring good strength and transparency. It has fairly good resistance to moisture vapor transmittance.

**polyethylene** — a plastic resin (polymer of ethylene) that has found very many applications in packaging, both as a film and for bottles. It comes in high and low density varieties, oriented and non-oriented, and with other modifications; also available as films combined with many other materials.

**polyglycerol ester** — emulsifiers produced by first polymerizing glycerol and then esterifying the polymer with selected fatty acids.

**polymerization** — (1) A chemical combining of two or more molecules of the same type, leading to long chains that may or may not be branched and that have qualities very different from the monomer. (2) A deterioration in frying fat resulting from the combining of fatty acids to give gum-like substances that cause foaming and other undesirable phenomena.

**polymorphism** — having the potential of existing in more than one crystalline form. Many fats can exist in at least three different crystalline forms — alpha, beta, and beta prime. The crystal form of a fat has an important effect on its melting point and its performance in use.

**polvorón** — *Spain* hazelnut cookie.

**polypropylene** — a plastic resin formed of polymerized propylene. Used in many of the same ways as polyethylene, but PP is generally tougher, harder, and (in some forms) more transparent than PE. Much used in packaging films.

**polysaccharide** — a polymer of several sugar molecules, such as starch, which is a polymer of glucose.

**polysorbate 60** — commonly known as Tween, this emulsifier is used in cakes and adjuncts to give greater volume, and in bread to retard staling and strengthen doughs.

**polystyrene** — a thermoplastic resin consisting of polymerized styrene. It is used primarily in rigid and semi-rigid items rather than in films. Has excellent transparency and is cheap.

**polythene** — *UK* polyethylene.

**polyunsaturated fats** — triglycerides containing one or more polyunsaturated fatty acids.

**polyunsaturated fatty acids** — fatty acids containing two or more double bonds joining carbon atoms.

**polyvinyl chloride** — a plastic resin used mostly for making packaging films; has good grease and solvent resistance, but not particularly good moisture resistance.

**polyvinylidene chloride** — a plastic film tradenamed Saran. Very good strength and low moisture vapor transmission.

**pomme** — *France* apple.

**pomme (de terre) chips** — *France* potato chips.

**ponchiki** — *Russia* fritter.

**poolish** — *France* a baking intermediate similar to sourdough starter.

**poori** — also, "puri." *India* a chapati that has been fried in deep fat.

**popcorn** — a specific variety or type of corn characterized by having relatively small, hard, spheroidal kernels that rapidly expand into a white, soft, irregular mass (i.e., they "pop") when subjected to a rapid influx of heat.

**popovers** — a quick bread made from a thin batter rich in egg, which, when rapidly heated, expands so as to give a very open and very coarse-grained interior. The expanded balls are then dried at lower temperature.

**poppyseed** — small, kidney-shaped, bluish-black seeds with a nutty flavor. Used as whole seeds in cakes and for scattering over the tops of bread rolls, and (when ground) as the basis for fillings in kolache and other European pastries.

**porgandi plaadipirukas** — *Estonia* open-faced carrot pie, made with a yeast-leavened dough.

**porosity** — (1) Average size or size distribution of the cells in the crumb of a loaf of bread; described as open (large cells) or close (small cells). (2) The extent to which unusually large cells are found in the crumb of bakery foods; this kind of "porosity" is desirable in English muffins but not in most other bakery foods.

**porridge** — a food made by boiling some farinaceous or leguminous substance in water or milk, so as to form a thick soup or thin pudding, e.g., a cooked breakfast cereal, such as oatmeal.

**porter** — *UK* an alcoholic beverage similar to beer, but dark in color (due to use of roasted malt in the mash), relatively sweet, and moderate in ethanol content. Essentially a weak stout.

**Portugal cakes** — *Old English* a sweet rich dough, leavened only with beaten whole eggs, flavored with rosewater and wine, containing currants, and baked as small cakes or muffins.

**positive displacement meter** — a volumetric fluid-measuring device that relies upon the repetitive filling and emptying of a cavity of known size by a piston.

**positive displacement pump** — a pump that uses a reciprocating or rotating cavity to deliver a predictable volume of fluid at each complete cycle.

**post-mixer developer** — a machine of the mixing or extruding type that subjects a mixed and partly fermented dough to kneading action for the purpose of removing gas bubbles so that dividing can be more accurate and, possibly, for improving the physical status of the dough.

**Postum** — trade name for a flaky-powdery preparation of (primarily) cereal ingredients, used as a caffeine-free substitute for instant coffee.

**potable** — suitable for drinking by human beings.

**potassium bicarbonate** — a carbon dioxide source that has been suggested for replacing sodium bicarbonate in low-sodium baked goods.

**potassium bitartrate** — also, "cream of tartar," and "acid potassium tartrate." The original acid-reacting substance used in chemically leavened goods; obtained from the lees in wine vats or synthesized.

**potassium bromate** — a chemical substance ($KBrO_3$) used as an oxidizer in the baking industry. It is relatively slow acting.

**potassium chloride** — used as a substitute for salt (sodium chloride) in foods intended to appeal to those persons who believe the latter substance is deleterious to their health.

**potassium iodate** — a chemical used to oxidize gluten proteins to make extensible doughs appear stronger and drier. It generally reacts more rapidly than potassium bromate.

## GLOSSARY OF CEREAL TECHNOLOGY TERMS 153

**potassium sorbate** — a preservative that acts similarly to sorbic acid but is often easier to use because of its greater solubility.

**potato bread** — generally, yeast-leavened white bread to the dough of which has been added a generous portion of either mashed cooked potatoes or (usually) dehydrated potato flour.

**potato flour** — a powdered material prepared by dehydrating cooked peeled potatoes, often with additives to improve performance or flavor.

**poulard wheat** — a variety of *Triticum turgidum* that is rarely grown in the US; suitable only for stock feed; its yellowish to red kernels are short, thick, and humped.

**pound cake** — in modern terminology, this name is applied to a firm yellowish cake of fine cell structure, rather dense, and often in loaf form; the flavor is usually mild lemon or vanilla. The traditional formula is a pound each of flour, sugar, butter, and liquid whole eggs, but many different formulas are being used today.

**pouring icing** — a type of glaze sufficiently low in viscosity to permit it to be applied by pouring over a cake, cookie, doughnut, etc. Often contains only a few ingredients, e.g., powdered sugar, water, salt, flavor, and color, with perhaps corn syrup or a stabilizer.

**powdered sugar** — cane or beet sugar that has been very finely ground; it is available in different particle sizes. Commercial products contain about 3% corn starch and occasionally small amounts of other additives to prevent the formation of lumps due to moisture absorption. Synonyms: confectioners' sugar, icing sugar.

**powder feeders** — dispensers designed to deposit fine powders, usually based on a swept metal sieve below a sometimes vibrated hopper.

**pozharske** — *Russia* a type of savory pasty.

**praline** — A filling or flavoring for chocolates and the like that is made by roasting almonds (or other nuts) in molten sugar until they are brown and crisp, and then grinding the cooled "brittle."

**pralines** — filled chocolate candies in the European style.

**pratie oaten** — *Ireland* a mixture of mashed cooked potatoes, oatmeal, and salt that is cut or formed into pieces, and cooked on a griddle.

**precipitate** — a solid material that separates from a solution as the consequence of some chemical or physical change.

**pre-ferments** — basically the same thing as a water broth, containing typically yeast, sugar, yeast food, vitamins, mold inhibitor and, perhaps, a dairy product. Fermented for about 2 to 2.5 hr before adding to the other ingredients in the sponge or dough stage.

**preheat** — allow an oven to reach, and stabilize at, the desired baking temperature preparatory to placing dough in the oven.

**premix** — (1) A combination of ingredients, uniformly blended, to which is to be added water and possibly other ingredients to make a finished dough or batter. Advantages can be: fewer total measuring steps, especially for minor ingredients, and greater accuracy. (2) Also, see "prepared flour mixes."

**premixed gas system** — a fuel delivery method in which the natural gas (or the like) is mixed with air in a separate step before it enters the burners.

**prepared flour mixes** — also, "prepared bakery mixes" and "prepared baking mixes," Blends of flours with other ingredients that are intended to reduce the measuring steps required at the point of mixing.

**preprzen hleb** — *Serbia/Croatia* toast.

**pre-scoring** — cutting English muffins horizontally at their sides from their edges about half way to the middle to make it easier for the consumer to split the muffin into halves.

**preservatives** — materials added to foods for the specific purpose of increasing shelf-life; the term is usually applied to chemicals that prevent or retard microbiological growth (e.g., sorbic acid).

**preserves** — a cooked sugar and fruit mixture similar to a jam but including relatively large pieces of fruit.

**pressed crumb crusts** — a mixture of cookie or cake crumbs with shortening, and, usually, sugar that has been pressed in a baking pan (or a die shaped like a baking pan) to give a crust suitable for filling with a pre-cooked pudding-like mixture.

**press out** — to divide a batch of dough into a specified number of pieces by means of a dough-press (bun press, roll press) machine.

**pressure depositor** — a device for extruding doughnut dough and the like from a pressurized chamber.

**pretzels** — formerly applied only to crisp snack items with a unique shape, formed from yeast-leavened dough and dipped in an alkali solution before baking. Now, also applied to soft products of the same general shape and to crisp products made in the form of sticks, nuggets, and fancy shapes.

**pretzel salt** — sodium chloride in rather large particle size, sometimes in flake shape, used to coat pretzels.

**prinsesstårta** — *Sweden* sponge cake layered with vanilla custard and whipped cream and covered with green almond paste.

**printe** — *Germany* a kind of honey-flavored cookie.

**process flavors** — flavored (and colored) compounds originating mostly from Maillard reactions occurring during the heat treatment of foods, e.g., roasting and frying.

**product stream** — any one of 125 to 150 mill streams in the usual flour manufacturing process.

**profiterole** — hollow ball of baked cream puff paste, usually filled with pastry cream; profiteroles of small size are used as the building blocks for constructing multiple-serving desserts of elaborate design.

**profiterole au chocolate** — *France* puff pastry shells filled with whipped cream or custard and covered with melted chocolate.
**proja** — *Serbia/Croatia* bread made from cornmeal.
**proof** — expansion of the dough from the time it is formed into a loaf (or other final shape) until it enters the oven, or the period during which this occurs.
**proof box** — a chamber or room equipped with means for maintaining relatively constant temperature and humidity and for transporting dough containers such as loaf pans through the space.
**proofing** — the final rising of bread dough (except for oven spring) occurring after the dough has been formed into the final piece; also applied to the final fermentation step of other yeast-leavened goods.
**proofing enclosures** — proof boxes.
**pro-oxidant** — a substance that promotes or catalyzes an oxidation reaction. For example, materials that accelerate the development of oxidative rancidity in a foodstuff containing fat.
**propionates** — salts of propionic acid used as preservatives; they are primarily effective against mold growth.
**propionic acid** — an organic acid compound based on a chain of three carbon atoms, known to have a growth-restricting effect on certain microorganisms.
**proportional belt feeder** — a measuring dispenser for ingredients based on the principle of timing the duration of delivery of a dry particulate material that has deposited in a uniform layer on a belt moving at a constant rate.
**propylene glycol esters** — lipophilic emulsifiers used to improve whipping response and volume in cake batters and whipped toppings.
**propyl gallate** — an antioxidant.
**protease** — proteinase.
**protected active dry yeast** — an ADY that contains antioxidants and is dried to a lower moisture than the regular form of ADY; as a result, it has a longer shelf life.
**proteinase** — an enzyme that splits protein molecules into peptides or amino acids.
**protein hydrolysates** — mixtures of peptides, amino acids, and modified proteins formed by the action of proteinases and/or acids on proteins, whether isolated or in crude mixtures. Much used as flavoring agents and sometimes as nutrition-enriching agents.
**protein quality tests** — analytical methods applied to a sample of, e.g., wheat to determine the suitability of its protein for baking or some other purpose.
**proteins** — large molecules composed of amino acids; proteins are ubiquitous in living tissues of plants and animals.

**protein strength** — a rather inexact term usually indicating the satisfactoriness for use in breadmaking of the gluten in a particular sample of wheat flour.

**proving** — same as proofing, the former spelling being more common in England.

**provitamin A** — a compound utilizable as vitamin A by the body, though not chemically identifiable as the vitamin.

**prugna secca** — *Italy* prune.

**pryaniki** — *Russia* gingerbread.

**PSIG** — gas pressure (in pounds per square inch) measured without reference to atmospheric pressure.

**psomi** — *Greece* bread.

**psomi imilefko** — *Greece* white bread made with unbleached flour.

**psomi khoriatiko** — *Greece* rye bread.

**psomi mavro** — *Greece* wholewheat bread.

**psomi me susami** — *Greece* bread with sesame seeds..

**psychrometer** — a device for measuring relative humidity.

**psychrophilic** — describes microorganisms that can grow and reproduce at relatively low temperatures, e.g., 40°F.

**pudding** — (1) A soft sweet (dessert) gelled product often texturized with a starchy material and flavored with almost anything. (2) *UK* Fairly lean doughs made up into multiple-serving pieces, say 1 lb or more, and cooked by boiling or steaming; though alleged to be a dessert-type product, the dough is not very sweet, but is enriched by a fatty material (traditionally suet) and raisins or the like; often served with jam or a sauce of custard formulation and consistency. Also, the term "pudding" is rather widely used to refer to the dessert course of a meal, regardless of its type.

**pudding cakes** — cakes made from a batter which, though uniformly mixed, separates into a relatively soft gelatinous mass and a relatively firm cake-like mass during baking.

**pudim** — *Portugal* pudding.

**pudim à portuguesa** — *Portugal* custard flavored with brandy and containing raisins.

**puf böregi** — *Turkey* chemically leavened puff pastries that are fried in shallow oil; generally made in small pieces; with various fillings, both savory and sweet.

**puff biscuits** — also, "puff crackers." Crackers of the Ritz or snack type, but relatively high in fat and slightly sweetened, and with the dough being processed by an interleaving method similar to the technique used for puff pastry so as to give more layering and flakiness in the finished product.

**puffed cereals** — cooked whole grains or pellets of gelatinized formulated doughs that have been highly aerated as a result of the rapid evolution of steam the interior of the particle when it moves from a hot pressurized container to normal atmospheric pressure conditions. See also, "oven-puffed."

**puff paste** — also, "puff pastry dough." An unleavened dough that has been interlayered with butter or other shortening, then rolled out, folded over and sheeted again. The folding and sheeting operation may be repeated several times. If properly made, these doughs yield flaky, tender, and glossy baked products.

**puff pastry margarine** — margarine specially made for layering puff pastry doughs; it is firmer than regular margarine because it contains fats of higher melting point, and it usually contains very little salt and not much water.

**puits d'amour** — *France* pastry shell filled with liqueur flavored custard.

**puliszka** — *Hungary* cornmeal dumplings.

**pulla** — *Finland* bun.

**pullman bread** — bread baked in a pan of (nearly) square cross-section having a cover. Sandwich bread.

**pullman pans** — long metal baking pans of almost square cross-section fitted with removable lids.

**pumpernickel** — in its original form, this is a very dense rye bread made entirely of crushed rye kernels with no admixture of wheat flour, and baked for a very long time (many hours) at a relatively low temperature so as to develop a caramelized flavor. Now applied to many other kinds of dark rye bread. One of the few types of rye bread that is not routinely flavored with caraway seeds.

**punching** — deflating, or pressing the excess gas out of a fermenting dough or sponge by kneading it. This is a process used in the manual preparation of doughs. In sponge-and-dough operations, the remix step accomplishes the same result.

**punc torta** — *Serbia/Croatia* frosted sponge cake sprinkled with rum.

**pup loaf** — a small (e.g., 100 to 150 g) loaf of bread prepared in a laboratory procedure to determine the baking quality of flour or other ingredient.

**pura** — *Serbia/Croatia* cornmeal mush.

**puri** — *India* chapati dough sheeted and deep-fried; small puffed breads fried in ghee.

**purifier** — a device in which an air current is forced up through a bed of middlings bouncing on an oscillating sieve so as to blow the bran out of the mixture and into a separate take-off tube.

**pusher bar loader** — a machine that uses a reciprocating metal bar to push straps of pans into an oven at the proper time to fill a tray or shelf that has appeared at the oven entrance.

**puter** — *Serbia/Croatia* butter.

**puuro** — *Finland* porridge.

**pyaz** — *India* onions.

**pyechyenye** — *Russia* cookies.
**pyechyonyj** — *Russia* baked.
**pyelmyeni** — *Russia* pocket of unleavened dough containing a filling such as cabbage or meat, cooked in soup but served separately; like ravioli.
**pyridoxine** — vitamin $B_6$, found in wheat germ and many other food materials.
**pyrometer** — measuring device designed to determine high temperatures.

## -Q-

**qualitative** — describes an evaluation or description that does not contain numerical measurements determined by actual physical tests.
**quality** — (1) An attribute or characteristic of something. (2) The extent to which an item or process approaches a standard of perfection.
**quality parameters** — subjective specifications or analytical ranges that define the relative goodness of a product or thing for its intended use.
**quantitative** — an evaluation or description that includes measurements, as of degree or amount (e.g., thermometric data, weight determinations).
**quart** — in the U.S.A., a quart is 32 fl oz; in the UK and Canada, an Imperial quart is 40 fl oz.
**queen of puddings** — *UK* a baked pudding topped with jam and meringue.
**queijada** — *Portugal* small cottage-cheese tart.
**queijada de Sintra** — *Portugal* small cottage-cheese tart flavored with cinnamon.
**quenelle** — *France* a light dumpling made of fish, fowl, or meat, often with a binder of flour or crumbs.
**quesadilla** — *Spain* cheesecake.
**queso** — *Spain* cheese.
**quiche** — a flan or open tart containing meat or vegetable filling, eggs, and cream. Many of these products are made in a crust very similar to a fruit pie crust, while others (especially in Europe) are placed on a yeast-leavened dough base. The filling is nearly always a non-sweet custard-like concoction including vegetables, meat pieces, etc.
**quickbreads** — bread- or roll-like products such as muffins, baking powder biscuits, scones, etc. Because they are chemically leavened doughs, the lengthy fermentation steps required for making regular bread and rolls are not needed; usually low in enriching agents, although date-nut breads and the like (actually cakes) are sometimes considered to be in this category.
**quick-cooking oats** — also, "quick oats." Oat groats that have been cut into several pieces and then rolled into thin flakes, so that the cooking time is substantially reduced, as compared to whole groats.
**quick-cooking rice** — whole-grain rice that has been processed, usually by partially precooking then drying, so as to develop an acceptable consistency when the consumer boils or steams them for a shorter time as compared to regular brown or milled rice.
**quick-freeze** — to freeze food so rapidly that the ice crystals formed therein are too small to seriously impair the cells or other microscopic structures in the matrix.
**quinoa** — pigweed, *Chemopodium quinoa*, or the seeds from this plant, used as a food in Peru and perhaps elsewhere.

## -R-

**rabanada** — *Portugal* a bread product similar to French toast.
**rack** — a metal frame for holding trays or pans in a vertical array; may be movable or stationary.
**rack cooler** — a cooling chamber, usually provided with circulating refrigerated air, that has provisions for taking in and pushing out racks of baked products or of doughs to be retarded.
**rack ovens** — ovens in which racks (usually on casters) containing the baking pans are pushed in and pulled out of the baking chamber either manually or (usually) mechanically.
**rack proofers** — a tier of shelves, usually of wire or rods, contained in a frame supported by casters, used for holding pans or trays of yeast-leavened loaves or rolls during the final fermentation (proofing) stage; these racks are moved into and out of humidity- and temperature-controlled chambers either manually or by automatic conveyors.
**radar ovens** — an early term for microwave cookers.
**radiant heat** — heat transferred by radiation.
**radiation** — the transfer of heat energy by electromagnetic rays, such as infrared rays.
**radiation cooling** — also, "radiant cooling." refrigerating, especially of chocolate confections, by conveying the food pieces through a tunnel containing chilled surfaces above the conveyor; i.e., cooling not dependent (at least, primarily) on convection or conduction.
**raffinose** — a crystalline, colorless, slightly sweet trisaccharide yielding fructose and melibiose upon hydrolysis. Small amounts found in sugarbeets, cottonseed, etc.
**råg** — *Sweden* rye.
**rågbröt** — *Sweden* rye bread.
**raghif** — *Israel* thick flat bread made from whole wheat flour and other ingredients. Traditionally baked on a bed of pebbles.
**ragi** — *India* seeds of the finger millet, *Eleusine coracana*.
**rahkapiirakka** — *Finland* a kind of cheesecake.
**raised** — essentially synonymous with "leavened."
**raisin bread** — yeast-leavened bread and rolls containing whole raisins distributed throughout the dough, the latter usually being slightly sweetened and often mildly spiced.
**raisin paste** — finely ground raisins, usually without added ingredients.
**raisins** — dried sweet grapes, normally dark brown or black, but may be bleached (usually with sulfur dioxide) so as to appear yellowish.
**ram** — the part of a dough divider that forces dough into the compression chamber.

**ramazan pidesi** — *Turkey* a yeast-leavened flat bread similar to nan.
**ramen** — *Japan* a kind of pasta, usually made of wheat flour or meal, but rarely of durum semolina, often in the shape of spaghetti or vermicelli.
**rån** — *Sweden* small wafer.
**rancidity** — a type of fat deterioration which causes unpleasant odors and/or tastes; can be of either the oxidative or the hydrolytic type.
**random-load grouper** — a machine that automatically assembles groups of pans or straps of unbaked dough (such as bread loaves) for placing on conveyors, its distinguishing characteristic being that it allows any size of group from one pan up to a full shelf to be loaded at one time.
**ranskanleipä** — *Finland* white bread.
**rántott borsó** — *Hungary* very small lumps of a dough made from egg, flour, milk, and salt, and fried in shallow oil; the dough is pressed through a sieve or extruded in some other manner directly into hot lard. Have been called "fried soup peas."
**rapeseed** — the grain from which canola oil is pressed; the small, round, dark-colored rapeseeds have been used as the measuring medium in volumetric displacement devices designed for determining the volume of loaves and other baked products.
**rapeseed oil** — the oil pressed from rapeseed, now called "canola" after certain processing steps have been applied.
**rarou** — *Egypt* a thin crisp bread that is consumed after moistening or is crumbled like crackers and eaten in soup, etc. The dough consists of a mixture of equal proportions of wheat flour and corn flour, 10% okra flour, salt, and water. After the dough ferments for about an hour, it is pressed into a thin layer and baked.
**ratafia biscuits** — *Old English* biscuits of the "Naples"type, originally flavored with ratafia liqueur (now rare or unobtainable) or sherry.
**rate of reaction** — for chemical leavening systems, the rate at which carbon dioxide is evolved under controlled conditions.
**ratio** — the quantity of two or more materials as related to one another.
**ravanie** — *Greece* cake leavened with whipped egg whites and containing ground walnuts, almonds, or other nuts, often containing zwieback crumbs, soaked with rum- or orange-flavored syrup after baking.
**raviole** — *Serbia/Croatia* ravioli.
**ravioles** — *Spain* ravioli.
**ravioli** — thin pasta sheets completely surrounding a filling of meat, cheese, etc. Usually boiled and served with a sauce, as of tomato.
**raw materials** — essentially all the materials used to make up a product; ingredients, even though some of them may be cooked, not "raw." The preferred term is "ingredients."
**raw sugar** — granular cane sugar that has had some of the final purification steps omitted; it is always darker and higher in ash, and usually

coarser, than granulated cane or beet sugar. This is not the brown sugar offered at retail.

**Reaumur** — a temperature scale in which 0°R is the freezing point of water and 80°R is the boiling point. A traditional scale for European brewers, not used much anymore.

**rebanadas** — *Mexico* sliced and toasted white bread spread with butter-flavored cream and powdered sugar.

**rebanar** — (v.) *Spain* to slice.

**rebozado** — *Spain* breaded or fried in batter.

**recipe** — a list of the ingredients and their quantity as needed for preparing a product; differs from "formula", if at all, in that recipes are denominated in household measurements and yield small quantities of product while formulas are in metric amounts and are used to prepare wholesale quantities (or lab quantities) of product. Most recipes also include processing instructions while formulas generally do not, though there is no hard and fast rule.

**reciprocal baking** — a method of allocating production capacity among a multi-plant bakery so that plant "A" will bake one type of product and plant "B" will bake a different type; the plants then ship items to regional centers for final assembly and distribution to retail outlets.

**reciprocating** — any mechanical assemblage that involves a back and forth movement as opposed to either rotation or continuous movement in one direction.

**reciprocating slicer** — equipment for slicing bread loaves that moves in an up-and-down motion saw-like blades set in a frame.

**reconstitute** — to rehydrate a dried material to more-or-less its original water content; sometimes, less accurately, used to mean thawing a frozen product.

**recovery periods** — the rest stages that allow dough pieces to accumulate gas and the gluten to relax following, for example, rounding.

**red dog** — not a carmine canine, but a very low-grade flour with a high content of bran. This by-product is not often used for bakery products, but with the current craze for high fiber products it could become a premium ingredient.

**red oats** — a species or type of oats grown in regions thought to be too warm for satisfactory production of the common oat.

**red pepper** — (1) Cayenne pepper or similar spices. (2) Edible fleshy peppers such as bell peppers that have a reddish color.

**red rice** — a species of rice (*Oryza rufipogon*) with red seed coat and pinkish grain, sometimes used as a food in India, but rarely seen in the US.

**reduced calorie** — describes a food having fewer calories than an equivalent amount of a comparable item of traditional composition; this description can be applied to foods only if strict Federal regulations are followed.

**reduced iron** — very small particles of metallic iron; has been used as a mineral supplement in bakery products and some other foods.

**reducing agent** — an ingredient such as glutathione or L-cysteine, that adds hydrogen to the disulfide bonds in gluten, resulting in a weakening of the protein network.

**reducing sugar** — a sugar that reduces Fehling's solution; related to the position of the aldehyde or ketone group in the molecule. Most sugars are reducing sugars, but sucrose is not.

**reductants** — a synonym of "reducing agents," for all practical purposes.

**reduction** — procedure by which grain is milled into flour and by-products.

**reduction rolls** — in the flour milling process, pairs of rotating steel cylinders that grind particles of wheat endosperm into fine flour.

**reel** — a rotating, generally cylindrical, sieve, with no beaters inside and, in most cases, aligned at a slight angle to the horizontal.

**reel ovens** — early name for the revolving tray oven; now used mostly for the smaller type of revolving tray ovens, as found in some retail bakeries.

**refiners' syrups** — a group of sugar syrups including both refined sugar and partially refined sugars in variable proportions, most of the content of all of them being sucrose but with some invert sugar, and with more color and flavor than pure sucrose syrup.

**refractive index** — also, "RI." The refractive index of a substance is the numerical expression of the ratio of the speed of light in a vacuum to the speed of light in a test object. For practical measurements, the scales of instruments show the refractive indexes compared to air rather than to vacuum. Simple, relatively inexpensive instruments utilizing this principle are often used to determine the so-called soluble solids in certain kinds of foodstuffs, e.g., sugar syrups, corn syrups, and fruit juices, and to identify and characterize fats and oils.

**refrigerated dough products** — a term that has been applied mostly to chemically leavened doughs, usually partially formed into biscuits and the like, that are distributed in compressed form in composite tubes; such products must be refrigerated during storage and distribution if adequate quality is to be maintained for the two or three months of normally expected shelf-life.

**refrigerants** — (1) Materials, such as liquid nitrogen, which can be used to cool products. (2) Heat transfer media, such as ammonia, used in refrigeration systems, including freezers.

**reibekuchen** — *Germany* potato pancake.

**reikäleipä** — *Finland* rye bread prepared as a ring-shaped loaf.

**reis** — *Germany* rice.

**relative humidity** — the ratio between the amount of water vapor actually in a sample of air, and the greatest amount of water vapor the same air could contain, both being at the same temperature and pressure.

**release** — in flour milling, the amount or percentage of stock yielded by a break step and which does not go on to the next break step, but is sent to purifiers, etc., for grading.

**rellenar** — (v.) *Mexico* to stuff or fill, e.g., "rellenar las empanadas," to put filling the pies.

**relleno de frutas** — *Mexico* fruit filling.

**remix** — in the sponge-and-dough process of breadmaking, the operation of mixing the fermented sponge with the rest of the ingredients.

**remonce** — *Denmark* pastry filling, such as butter cream.

**rendering** — application of heat to a fatty animal tissue in order to release the fat for collection or further processing.

**reposteria** — *Mexico* these pastries are similar to butter cookies in taste and texture, and are often finished with fruit toppings or sprinkles; chemically leavened. powder.

**residual chlorine** — the amount of chlorine over and above the amount needed to disinfect a potable water supply; it is useful in eliminating contaminants that may enter the water during its flow through the pipes of the distribution system.

**residual sugars** — sugars that remain in a finished bakery product, i.e., sugars that have not been fermented.

**response** — the reaction of a dough to the set of conditions it encounters during preparation.

**rest period** — the time given a dough to recover desired properties after some processing step has been applied.

**retailer ovens** — baking ovens of a size and complexity suiting them for use in retail bakeries, in-store operations, restaurants, etc. Deck ovens and rack ovens are two types meeting these requirements.

**retarder** — a refrigerator used for retarding doughs.

**retarding** — delaying fermentation of dough by chilling (not freezing) it.

**rétes** — *Hungary* strudel.

**rétestészta** — *Hungary* strudel dough.

**retrogradation** — change of a cooked starch from a gel to an insoluble state, said to be due to formation of aggregates of the starch molecules.

**returns** — (1) Mill stock that is sent back for retreatment to the same machine or even to a machine nearer the head of the mill. (2) Finished products that have been sent back to the manufacturer by a wholesaler, retailer, or consumer because the products are outdated or have other unacceptable characteristics.

**revani** — *Turkey* semolina mixed with butter, sugar, and eggs, then baked and, finally, soaked with slightly sweetened fruit juice. *Greece* A kind of sponge cake soaked in syrup.

**reverse molding** — in the mechanical forming of a bread loaf, a process that reverses the sheeted dough piece so that the formerly leading end enters the curling section of the molder last. Supposed to improve texture in the baked loaf.

**reverse osmosis** — a process for removing dissolved molecules from a liquid (usually water), based on the application of pressure to force the liquid through a semipermeable membrane, against the normal gradient of osmotic pressure.

**reverse sheeting molder** — a bread loaf-forming machine utilizing the principles of reverse molding (which see).

**reversible sheeters** — batch-type dough sheeters in which the direction of travel of the input and take-away conveyors, and the direction of rotation of the sheeting rollers can be reversed in a coordinated fashion so that a batch of dough can be repeatedly thinned without manual intervention.

**reversion** — an undesirable change in the flavor of a refined oil or fat, so-called because it was thought to result when a refined, nearly flavorless oil reverted to its original "raw" flavor.

**revolving oven** — a reel oven.

**rezanci** — *Serbia/Croatia* noodles.

**rheology** — the science concerned with the deformation and flow of matter.

**Rheon encruster** — a machine for wrapping dough (or other materials of similar rheology) completely around a single portion of a filling having a quite different consistency; operates in a continuous manner but forms only one dumpling at a time.

**ribes** — *Italy* currants.

**ribbon blenders** — mixers consisting of a large trough or drum, horizontally aligned, within which rotate narrow strips of metal connected by spindles to a central axle. Found in many sizes and designs, and particularly suitable for blending powders.

**riboflavin** — vitamin $B_2$, yellowish-green in color when purified; required to be added to enriched flour.

**ribs** — *Denmark* currant.

**rice** — the grain obtained from the plant *Oryza sativa*. Has a pleasant mellow flavor, but contains no gluten so that an extensible dough cannot be made from it. Commercially available forms are rough rice (with the hulls on), brown rice (not polished, but hulls removed), and polished rice (the usual form of commercial rice). Wild rice is an entirely different grain.

**rice bran oil** — the oil solvent-extracted from the outer layers of rice that are removed in the milling process.

**rice cake** — a moist product made from milled rice that has been soaked, drained, kneaded, formed, packed in sealed plastic film, pasteurized, and cooled. A convenience food for consumers who do not wish to boil rice.

**rice cakes** — dry puffed rice shaped into thick disks, usually after the addition of various condiments and adhesives, intended to be consumed without further preparation.

**rice crackers** — a type of snack cracker popular in parts of Asia, resembling somewhat a savory biscuit but they contain no wheat flour.

**rice flour** — finely ground rice kernels; used as an ingredient and as dusting powder.

**rice milk** — a white to beige-colored slurry having a distinctly sweet, rice-like flavor; made from cooked rice incubated with koji, then the liquid ("milk") separated by filtration.

**rice noodles** — shapes similar to linguini or spaghetti made from a paste of steamed, ground non-glutinous rice. f

**rice paper** — (1) An edible, white, paper-like material made from the pith of a tree, *Tetrapanax papyriferum*, grown in China. (2) A thin paper made from rice straw.

**rice polishing** — coating fully milled rice kernels with a talc and glucose slurry, then drying in rotating drums. Purposes: make the rice whiter and glossier.

**rice ribbon noodles** — made in much the same way as rice stick noodles, but usually sold fresh for prompt cooking in the home.

**rice starch** — starch separated from rice grains; it has slightly different properties than corn starch. This term is often applied to rice powder, which is actually finely ground white that contains other substances (such as proteins) in addition to starch.

**rice stick noodles** — *Asia* short, flat noodles made by steaming a batter of rice flour and water in tubs, then brushing the cooked doughy mixture with oil, cutting it into thin strips of different widths, and drying.

**ricet** — *Serbia/Croatia* barley groats boiled with beans.

**rice wine** — a fermented beverage made from a mash of glutinous rice and starters containing a mixture of microorganisms. A typical example is the sake of Japan, but there are many other types made throughout Asia.

**rich doughs** — doughs that contain considerable amounts of some or all of the enriching ingredients, such as shortening, milk, sugar, etc.

**rieska** — *Finland* barley bread, unleavened.

**rigatoni** — *Italy* type of pasta similar to cannelloni, except rigatoni has longitudinal ridges on the outside of the tube.

**riisi** — *Finland* rice.

**rijst** — *Netherlands* rice.

**ripe dough** — a dough that has reached a stage of fermentation and conditioning making it suitable for make-up.

**ripeness** — readiness of dough for baking.

**ris** — *Sweden Russia Denmark* rice.

**risbrot** — *Denmark* rice bread, yeast-leavened, containing a large proportion of wheat flour as well as some cooked rice.

**rise** — the increase in height (or, less often, in volume) attained by a piece of dough in a given fermentation time.

**rise time** — in frying doughnuts and the like, the time required for the product to come to the surface of the hot fat after the raw dough piece has been dropped from the cutter.

**riso** — *Italy* rice.

**risotto** — *Italy* generally refers to a savory dish consisting of firm grains of rice that have been cooked in meat stock, then mixed with grated cheese, sour cream, etc. There are sweet dessert versions.

**rissol** — *Portugal* fritter containing ground meat or fish.

**rissole** — *France* fritter, pasty.

**riz** — *France* rice.

**riza** — *Serbia/Croatia* rice.

**rizi** — *Greece* rice.

**robotics** — the study and/or application of robots.

**robots** — programmable, multi-functional machines designed to move materials, parts, tools, or specialized devices through predetermined patterns for various purposes; devices that can perform a sequence of desired movements in response to mechanical, electrical, or electronic commands.

**rock bun** — *UK* slightly sweet roll or bun containing currants or raisins.

**rocks** — small, rough-surfaced fruited cookies baked from a stiff batter.

**rock salt** — salt from mines, used without any purification step involving dissolving but reduced in particle size to meet various needs. Flavor and appearance differ according to source.

**rock sugar** — very large crystals of sucrose.

**rogaliki orzechowe** — *Polish* walnut horns, an unleavened cookie containing a large proportion of ground walnuts and made up in crescent shapes.

**roggebrood** — *Netherlands* rye bread.

**roggenbrot** — *Germany* rye bread.

**rognures** — *Fr* scraps of leftover dough.

**rolat** — *Serbia/Croatia* roll-shaped cake.

**roles glass** — *Mexico* sweet rolls, topped with sugar glaze and usually containing a filling of cinnamon and/or raisins.

**roll dividers** — any machine that separates bulk dough into pieces of uniform size for later shaping into bread rolls.

**rolled pancakes** — *UK* a baked pancake, usually relatively thin and chemically leavened, rolled around a savory filling.

**roller-bed unloader** — a conveyor system for use at the exit port of ovens, consisting of a series of horizontal rotating tubes placed perpendicular to the pan movement.

**roller system** — or, "roller mill." The modern method of reducing wheat to flour, based on pairs of rotating cylinders between which grain and partially milled material are passed; a roller mill will also include many other devices, e.g., purifiers, sifters, cleaners, and dust collectors.

**roll-in doughs** — rich pastry doughs formed by repeated folding and sheeting after the sheets have been covered with a layer of fat.

**roll-in fat** — shortening specially processed to remain fairly stiff and cohesive during the making of puff pastry and other roll-in doughs.

**rolling-disc knives** — or, "rolling disc cutters." Blades of circular form, rotated on an axis by mechanical means, frequently used to cut strips from dough sheets moving beneath them. Often seen in arrays of several discs on a single axle.

**rolling pins** — wooden or metal cylinders, can be about 1 to about 4 inches in diameter and more than a foot long, usually with a handle at each end; employed for manually sheeting out dough.

**rolls** — portion-size pieces baked from lean yeast-leavened dough; the individual serving version of bread loaves; buns.

**rope** — a type of spoilage occurring in bread, now rare in the US. Characterized by a stringy, gummy degeneration of the crumb. *Bacillus mesentericus* is the responsible organism, its spores being sufficiently heat-resistant to persist through the baking process.

**roll winder** — a curling roller.

**roomboter** — *Netherlands* butter.

**roomli roti** — *India* a thin, crisp, flatbread.

**rope** — or, "rope spoilage," A type of spoilage occurring in bread, now rare in the US. Characterized by a stringy, gummy degeneration of the crumb. *Bacillus mesentericus* is the responsible organism, its spores being sufficiently heat-resistant to remain viable through the baking process.

**rosca de panque** — *Mexico* ring-shaped cake topped with powdered sugar.

**rosemary AR** — a spice extract that has been shown to have antioxidant properties.

**rosette** — (1) A bud-like ornament made from icing and somewhat resembling a rose. (2) A cookie or other flat disc with scalloped edges. (3) A type of cookie made by dipping a specially shaped metallic mold (having a handle connected to it by a rod) into a thin, slightly sweet batter, then immersing the coated mold into hot fat until the batter becomes crisp and somewhat browned. The thin shell removed from the mold may be finished with a dusting of cinnamon and sugar or, less commonly, coated with chocolate. The term is also used for the mold.

**rosin** — *Denmark* raisin.

**rosine** — *Germany* raisin.

**rosquilla** — *Spain* doughnut.

**rostat bröt** — *Sweden* toast.

**rotameter** — a fluid-flow measuring device, consisting of a vertical tube with a gradually tapered bore containing a float capable of unrestricted movement up and down the tube, the position of the tube indicating rate of flow.

**rotary cookie** — a cookie molded on rotary or Dutch equipment.
**rotary cutting machine** — dough forming machines that cut pieces from a sheet of dough by using a cylinder either embossed with sharp ridges or bearing metal strips bent into patterns.
**rotary molding machine** — or, "rotary machine." A cookie forming machine relying for its effects on a rotating metal or plastic cylinder having on its surface numerous shaped cavities into which dough is forced, later to be removed or ejected onto an oven band.
**rotary oven** — an oven having a horizontal rotating hearth that is circular in shape; now largely obsolete. The term is sometimes unsuitably used for revolving tray ovens.
**roti** — *India* bread.
**roti flour** — *India* a high extraction wheat flour of rather coarse granulation, widely used in Asia to make many types of unleavened bread, such as paratha and roti.
**roti jala** — *India* a wheat flour batter, usually including eggs and coconut milk, deposited from a perforated cup as interlacing strips onto a hot, ghee-coated pan. When the batter deposit is cooked on the bottom, it is turned over, cooked, then rolled up in a cylinder and served.
**roughage** — the insoluble, non-digestible material in food or feed; an older term, the meaning of which is about the same as "crude fiber."
**rouhesämpylä** — *Finland* whole wheat roll.
**roulade** — a layer of baked cake that has been spread with jam or some other kind of filling then rolled up; a jelly roll is one kind of roulade.
**rounder** — a machine that forms balls from roughly cut dough pieces preliminary to a fermentation step and/or additional forming operations.
**rounding** — rolling dough pieces coming from the divider so as to seal the surface and form a uniform skin on the pieces; can be either a hand or a machine operation.
**round top bread** — open top bread; loaves baked in pans that have no lids.
**rout press** — a bar press machine for extruding strips of cookie dough, usually it does not include a cutting device at the orifice.
**roux** — or "beurre roux." A uniform mixture of melted butter and flour (about equal proportions), slowly heated in a pan until it browns. Three kinds: white, blond, and brown, indicating increasing degrees of darkness. Used as a thickener and flavoring material for sauces.
**royal icing** — decorative frosting of cooked sugar and egg whites, notable for the speed at which it sets up.
**rugbrod** — *Denmark* yeast-leavened rye bread; several types are made.
**rugelach** — also, "rugula," "rugelah," and several other spellings. *Jewish* cookies made from a rolled out yeast-leavened dough containing cream cheese; a sugar-raisin-spice filling is placed on a dough sheet, which is rolled into a cylinder, sliced crosswise and then baked.

**ruisleipä** — *Finland* rye bread.
**rukkijahu** — *Estonia* rye flour.
**rulyet** — *Russia* pasty stuffed with meat or other filling.
**rum** — an alcoholic beverage distilled from fermented sugar cane juice. Occasionally used as a flavor in bakery dessert items (babas au rhum, etc.) and in confections; highly concentrated imitation rum flavors of satisfactory characteristics are available.
**rusk** — *Western Europe* a slice (usually circular) cut from a rich yeast-leavened baked dough (or one-half of a muffin-shaped piece) and re-baked until crisp and slightly browned. Both chemically leavened and extruded varieties are known.
**rye** — a grain that is darker and somewhat more fibrous than wheat and yields a flour that is stronger flavored but is much less effective in contributing uniform cellular structure to bread than is wheat flour.
**rye bread** — bread made from a blend of rye flour and wheat flour; rarely, it is made from all rye flour or crushed rye kernels, in which case a dense, moist, harsh-textured bread is obtained.
**rye sours** — sourdoughs incorporating a large percentage of rye flour or meal.

## -S-

**saccharify** — to convert into sugar; to impregnate with sugar.
**saccharifying enzyme** — beta-amylase; acts on starch molecules, causing the release of maltose residues from the end of glucose chains, a process that leads to the accumulation of limit dextrins.
**saccharin** — a chemical substance about 200 to 300 times as sweet as sugar, some consumers find it has a bitter flavor, as well; contributes no calories to the human diet.
**saccharine** — having the flavor quality of sugar; sweet.
**Saccharomyces** — a genus of microorganisms including several species of yeasts useful for fermenting foods and beverages.
**Saccharomyces cerevisiae** — taxonomical identification of bakers' yeast.
**Sacher torte** — also "sachertorte." A very rich confection consisting, in its traditional form, of a layer of baked chocolate cake (generally of rather dry texture) covered with apricot jam and enrobed with chocolate icing. Sometimes filled with a chocolate cream.
**sack cleaner** — a device for removing flour and other dust from the outside of sacked ingredients, either immediately after the sack is filled or just before the sack is dumped.
**Safe Drinking Water Act** — a federal law specifying maximum limits of contaminants allowed in potable water supplies.
**safflower oil** — a food oil of high unsaturation pressed from safflower seed.
**saffransbröt** — *Sweden* sweet saffron-flavored bread in roll or loaf form.
**saffron** — a spice consisting of the dried stigmas of the flower *Crocus sativus*. Used in saffron buns, and the like, to give an orange-yellow color and a unique aromatic/pungent flavor. Very expensive.
**saft** — *Germany Denmark* juice.
**saga-bolo** — *Japan* a Japanese confection made from a chemically leavened, very sweet dough. Sometimes, a mixture of buckwheat and wheat flours are used in this product.
**saggina** — *Italy* buckwheat.
**sago** — the starch separated from the pith of a tree, *Metroxylon sagu*. Pearl sago is formed by pressing the moist paste through a coarse sieve, the resulting balls being dried. Used as a basis for puddings and the like.
**sahne** — *Germany* cream.
**sajtos izelitö** — *Hungary* hot cheese crackers; a sort of savory sandwich cookie. The base cakes are disks of a baked unleavened dough of flour, butter, Swiss cheese, sour cream, and paprika; a paste of butter, cheese, and sour cream is placed between two baked discs.
**saka-manju** — *Japan* steamed Japanese bun made from fermented dough and containing bean jam filling.

**sakhar** — *Russia* sugar.
**sakhar naya pudra** — *Russia* powdered sugar.
**sakhar niy pyesok** — *Russia* granulated sugar.
**sakhar rafinad** — *Russia* refined sugar.
**sal** — *Spain* salt.
**salado** — *Spain* salted, salty.
**salad oil** — a refined, bleached, and deodorized edible oil that will, in most cases, also have been winterized. Some vegetable oils are natural salad oils, i.e., they do not form fat crystals at refrigerator temperatures.
**sale** — *Italy* salt.
**salim** — *Thailand* a dessert of thin tapioca noodles in a sauce made of coconut milk and palm-sugar syrup.
**Sally Lunn** — an old type of rich bread in muffin form that originated in England. Many formulas are extant.
**Salmonella** — a genus of aerobic Gram-negative bacteria that can cause food poisoning; they are especially common in improperly cleaned poultry products, but are and found in many other foodstuffs.
**salt** — unless the word is modified, to workers with food "salt" means common table salt, sodium chloride, which is commercially available in a wide range of purities and particle sizes and shapes.
**salt dispensers** — or, "salt spreaders." Machines that fit over a belt or other conveyor that carries a food product and deposit salt on the top of the food at a predetermined rate. There are three basic types of salt dispensers: mechanical, pneumatic, and electrostatic.
**saltines** — soda crackers topped with a small amount of coarse salt; the most common type of cracker.
**salt-rising bread** — a type of sourdough bread having a pungent aroma and dense crumb, the dough for which is ripened partly by a kind of salt-tolerant yeast and partly by bacteria.
**saltstaenger** — *Denmark* salted finger rolls, salt sticks.
**salt sticks** — cylindrical bread rolls with soft crumb (not crisp) and rather heavily coated with coarse salt crystals.
**salz** — *Germany* salt.
**salzburger knockerl** — *Germany* butter-fried dumplings made from beaten egg yolks, egg whites, sugar, and flour.
**sam** — *Serbia/Croatia* meringue.
**samosa** — also, "samoosa." *India* deep-fried triangular dumplings stuffed with potatoes or meat.
**sampita** — *Serbia/Croatia* whipped cream pie.
**sämpylä** — *Finland* roll.
**samrolna** — *Serbia/Croatia* whipped cream roll.
**sanding** — applying coarse sugar crystals to the surface of confections or bakery products.

# GLOSSARY OF CEREAL TECHNOLOGY TERMS 173

**sandiyi** — *Greece* whipped cream.
**sandkuchen** — *Germany* a baked product of the pound cake type, but with about half the wheat flour replaced by wheat starch.
**sandwich cookie** — a confection consisting of two baked base cakes (chemically leavened cookies) with a smooth creamy filling between them.
**sandwiching machine** — a device for combining base cakes and filling to make sandwich cookies.
**sandwich loaves** — pullman loaves, bread loaves with square cross-section resulting from being baked in special pans with lids.
**San Francisco sourdough bread** — white breads of strong flavor and odor resulting from often complex sour fermentations, a product allegedly having originated in San Francisco, CA.
**sangak** — *Iran* sourdough flat bread made from flour water, salt, and starter; sometimes sprinkled with sesame or poppy seeds. Traditionally baked on a hearth covered with pebbles.
**sanitarian** — a person responsible for evaluating, conducting, or directing sanitation procedures in a plant.
**sanitation** — the methods and practices used to prevent contact of foodstuffs with contaminants originating from the environment.
**sanitation audits** — a survey or inspection of premises, equipment, and environmental factors in and around a food plant that could cause biological contamination of products, always followed by a written report and often conducted by an outside specialist.
**santara** — *India* orange (the fruit).
**sapin sapin** — *Philippines* dessert made of rice flour cooked with coconut milk and sugar, deposited in layers of different colors and flavors.
**saponification value** — also, "SAP value." One of the tests performed to characterize edible fats and oils. Saponification is the hydrolysis of glycerides (such as fats) with an alkali to form free glycerol and fatty acids in the form of soaps. A reaction mixture is titrated with standardized alkali to get the SAP value, which is inversely related to the average molecular weight of the fat and, therefore, is an indication of the type of fatty acids in the fat. The larger the fatty acid molecule, the lower the SAP value.
**Saran** — a trade name applied to a group of tough, flexible thermoplastics composed mostly of polymerized vinylidene chloride and its copolymers.
**sarazin** — *Fr* buckwheat.
**saturated steam** — steam containing the maximum amount of water that can be held as individual water molecules (gas) at that particular temperature and pressure.
**saum** — *Serbia/Croatia* meringue.
**sausage roll** — puff pastry rolled around a cooked, ground meat mixture containing spices, then baked.
**savarin** — brioche dough baked in a ring mold, soaked with rum-flavored

syrup, and garnished with (or containing) almonds or other nuts, candied fruits, cream, etc. Several variations can be found.

**savory** — *UK* a highly-seasoned, not sweet, course served at the termination of a meal, e.g., an anchovy canape. More generally, and as used in the definitions in this Glossary, a flavor or food that is more closely related to the "salty" or main dish (meat/cheese/fish/nut) group than to the dessert (sweet/fruit/chocolate) type.

**scale** — (1) Any kind of weighing device. (2) To cut dough into pieces of the desired size (by weight or volume). (3) A linear arrangement of divisions in a particular measurement system, as the Celsius temperature scale.

**scaling** — (1) Cutting or otherwise separating a mass of dough into smaller pieces prior to further processing. The pieces will be uniform in weight but not necessarily of the same weight as the finished item, since some products require the assembling of one or more pieces to yield the shape and size desired. The dough piece may decrease in weight due to loss of moisture and other volatiles, and may gain in weight due to the acquisition of dusting flour, adjuncts, etc. (2) Weighing out ingredients.

**scalp** — to sift, as when separating the coarse part of the mill grind from the finer particles. Also, the removal of foreign material from the grain as it is being cleaned before milling.

**Schaal test** — the Schaal oven method for determining the oxidative stability of a fat-containing food involves heating a given amount of sample in a covered glass container at an elevated temperature until a rancid odor is detected. Results are reported as the length of time elapsing until the end point is reached. Temperatures, sample size, container type, etc., vary from analyst to analyst and should be specified when test results are recorded. This test can be quite useful if properly conducted and interpreted.

**schaumrolle** — *Germany* puff pastry shells usually filled with whipped cream or, sometimes, custard.

**schedule** — or "shop schedule." A sheet, form, or computer program used in the bakeshop to record the types and amounts of products to be made and the sequence and timing of their manufacture. More broadly, any kind of plan based on a listing of either times or a sequence of steps that is to be followed in a relatively complex operation.

**scheibe** — *Germany* a slice.

**schillerlocke** — *Germany* pastry cornet with a vanilla cream filling.

**schlagobers** — *Germany* also, "schlagrahm" and "schlagsahne." Whipped cream.

**schmier** — also, "shmear," "smear," etc. pan dressing.

**schnecke** — *Germany* cinnamon roll.

**schokolade** — *Germany* chocolate.

**schwartzbrot** — *Germany* a very dark, dense, moist type of rye bread; a name that is sometimes applied to types of pumpernickel.

## GLOSSARY OF CEREAL TECHNOLOGY TERMS 175

**schwarzwälder kirschtorte** — *Germany* a chocolate layer cake filled with cream and cherries, flavored with kirsch liqueur.

**scone** — a class or type of individual-serving hot breads that includes many varieties. Scones originated in Scotland as oatmeal-based cakes, but now virtually all of them are made of wheat flour and are chemically leavened. They are used in very much the same way as the baking powder biscuits popular in the southern U.S., but scones are usually a little richer and may include raisins. Most varieties are baked, but some are cooked on a griddle.

**scoop** — a shovel or ladle with rounded end and short handle, sometimes calibrated for volume; used for transferring dry ingredients to a scale pan or mixer bowl.

**scorch** — to overcook the surface of a product (especially with an open flame) so that the crust darkens excessively.

**scorched particle test** — a determination of the amount of discolored particles in dried milk, involving reconstitution and filtration steps.

**scoring** — (1) Judging finished goods according to descriptive scales of quality factors. (2) Cutting or slashing the top surfaces of dough pieces to give a characteristic pattern to the baked surface.

**Scotch bun** — see "black bun."

**scrap** — a product or an intermediate, such as dough, that is unusable in its existing form because of some defect and must be reworked or discarded.

**scrap meal** — out-of-standard but edible cookies that have been ground for use as a filler ingredient in highly flavored cookies. Usually the light- and dark-colored cookies are made into separate batches of scrap meal, i.e., light meal and dark meal.

**scratch** — as in "from scratch;" to form a dough, batter, etc., by weighing and combining all the individual ingredients in the bakery just prior to baking, as contrasted with the use of pre-mixes.

**scratch rolls** — in English usage, finely fluted mill rolls whose function is to remove bran fractions from the floury middlings.

**scratch system** — in some flour mills, a part of the roll system that is kept in reserve for use only when good release of the bran is not being obtained on the principal part of the system.

**screens** — (1) Sieve elements formed of woven filaments or perforated sheets having apertures of uniform size and shape; also, the sieves themselves. (2) baking pans for hearth breads consisting of wire-mesh forms mounted in supporting frames. (3) Flat wire-mesh rectangular supports on which doughnuts are proofed and then placed in the fryer.

**scroll** — a flake buster, q.v.

**sealing process** — an operation that causes the outer edge of a dough sheet that has been formed into a cylindrical loaf shape to be firmly adhered to the underlying dough.

**seam** — the outer end of the dough sheet used to form a loaf; visible in the unbaked loaf as a slight indentation bordered by a ridge down the side of the dough cylinder.

**seasoning** — the addition of spices, salt, etc. to foods. Also, these flavors.

**seb** — *India* apple.

**secer** — *Serbia/Croatia* sugar.

**secer u kocki** — *Serbia/Croatia* lump sugar.

**secer u kristalu** — *Serbia/Croatia* granulated sugar.

**secer u prahu** — *Serbia/Croatia* powdered sugar.

**seco** — *Portugal* dry, dried.

**sekacz** — *Poland* log cake, the dough being baked on a spit rotated in front of an open fire.

**sedimentation test** — a test for evaluating wheat protein quality — flour is suspended in an aqueous solution of lactic acid and held for a time under specified conditions, then the volume occupied by sediment is measured.

**seed dispensers** — machines that deposit sesame seeds, poppyseeds, or the like onto unbaked rolls or loaves passing under the dispenser.

**self-rising flour** — flour to which has been added sufficient sodium bicarbonate and leavening acids so that the consumer does not have to add baking powder to the recipe.

**sembei** — or o-sembei *Japan* rice wafers flavored with soy sauce or other seasonings. Rice and/or wheat flour is kneaded to make a dough, mixed with egg yolk, seaweed, spices, etc., then baked in a thin metal mold.

**semmelbrösel** — *Germany* breadcrumbs.

**semmelknödel** — *Germany* dumpling made mainly of diced white bread.

**sémola** — *Spain* semolina. *Portugal* "sêmola."

**semolina** — (1) Large middlings, passing through screens of 18 to 40 meshes per inch. (2) The purified middlings of durum wheat, used to manufacture the best quality of pasta.

**senbei** — *Japan* the generic name for Japanese crackers.

**senesen** — *Egypt* fiti (q.v.) made from a batter containing sorghum flour instead of wheat flour.

**sensors** — mechanical, electrical, or electronic devices that detect some sort of change in their environment (e.g., in the color of a cookie on a production line); sensors are almost always connected to some sort of computerized system that evaluates and reports their output.

**sepikujahu** — *Estonia* wheat flour.

**serowiec** — *Poland* also, "sernik." cheesecake.

**sesame seeds** — small flat white seeds of the sesame plant; used for topping bread rolls. A cooking oil is also obtained from these seeds but it is not much used in Western cuisines. There is also an edible black variety.

**sesame paste** — also, "sesame butter." A firm, slightly granular mass made by grinding sesame seed; there are two varieties. distinguished by whether or not the seeds have been toasted before grinding.

**setsuuri** — *Finland* sweetened rye bread.
**set up** — become firmer with time, or upon chilling, as gelatin desserts, some icings and puddings, etc.
**sevian** — *India* baked vermicelli in a creamy rosewater-flavored syrup.
**sfogliatelle** — *Italy* puff pastry with custard or jam filling.
**sfoliata zimi** — *Greece* kind of crust for pastries; puff pastry dough.
**shallots** — a member of the onion family, of somewhat milder flavor than most types of onions.
**shamsy** — *Egypt* rectangular- or disk-shaped bread with a light brown crust and a firm white crumb, made from a flour-water-salt dough. After the dough is formed into pieces, it is allowed to sit in strong sunlight for about three hours.
**shamy** — or "shamey." Yeast-leavened Egyptian flat bread made from 72% extraction flour. Baked so as to yield a "pocket," as in pita bread.
**sharps** — middlings.
**sharp-to-sharp** — the operation of roller mills when they are run so that the shorter face of the corrugated cutting edge on the faster rotating roller meets the longer face of the corrugations on the slower rotating roller.
**shau mai** — *China* steamed pork dumplings.
**sheen** — reflection or glossiness of the cell walls on a cut surface of a loaf; more generally, a shininess or matte gloss on the surface of any material.
**sheeter-laminater** — a machine found in most soda cracker production lines, but also in other plants, that forms a continuous dough strip by sheeting rollers, then folds the sheet into a layered structure that is thinned by rollers. The folding and sheeting may be repeated.
**sheeters** — machines that pass a dough piece or strip through the gap between a pair of metal rollers for the purpose of forming a dough sheet of uniform thickness.
**sheet pan** — a flat pan, often about 18x26 inches, with edges raised about an inch. Used for baking cookies, layer cakes, buns, hearth breads, etc.
**shelf life** — the length of time a product retains acceptable quality under average conditions of distribution and display.
**shell eggs** — eggs as they come from the hen, though usually cleaned, assorted by size, etc.
**shell top** — in bread, a condition where the top crust separates from the loaf, giving the appearance of a cap over the loaf.
**shingara** — *India* vegetable pasties.
**shinmai** — *Japan* a rice variety that is harvested in autumn; has moist, tender, sweet grains that require less water for cooking than ordinary rice.
**ship biscuit** — hardtack, pilot bread. The original form was a thick round piece of dough formed mostly from flour and fat, possibly unleavened, and baked to a fairly low moisture content. Traditionally, made without salt.
**shipstuff** — an old name for low-grade flour of high bran content.

**shirataki** — *Japan* (1) Noodles formed by extruding a paste made from the root konnyaku (devil's tongue). (2) Buckwheat noodles flavored with tea.

**shoofly pie** — a brown sugar and molasses flavored cake that is baked in a pie crust.

**shortbread** — (1) Rich cookie made from butter, flour, and sugar with little or no liquid added. (2) Baking powder biscuit dough enriched with sugar and shortening, used as a basis for strawberry shortcake and the like.

**short break** — the condition that exists when a loaf exhibits minimal oven spring. The break and shred along the side is very narrow in such loaves.

**shortcake** — can refer to shortbread, but now is often used for a portion of any kind of cake (even angel food, which contains no shortening) that is served topped with a generous portion of fresh or sweetened fruit.

**shortening** — a fatty ingredient used for the purpose of imparting desired textural qualities (especially, shortness, tenderness, or flakiness) to baked foods; sometimes applied with doubtful appropriateness to frying fats and oils. Shortenings may range in composition from pure natural oils to fairly complex mixtures of modified fats and additives.

**shortening value** — the extent to which a fat promotes tenderness and flakiness in pie crusts. Measured by different methods, usually by a subjective comparison of baked crusts.

**short-flake crust** — describes a baked pie crust that, when broken, discloses a structure of relatively small flakes.

**short-grain rice** — in addition to being relatively shorter than medium- and long-grain rice, these varieties (when cooked) usually seem moister, stickier, and softer.

**short paste** — a mixture of flour, fat, sugar, and water in various proportions. Generally much less elastic than bread dough.

**short patent** — a patent flour containing a relatively small percentage of the best flour streams; would not ordinarily be used to describe patent flour containing more than about 80% of the total flour output.

**shorts** — a low-grade mill product, containing principally germ and fine bran particles; used for animal feed.

**short system** — a milling process involving a relatively small number of reduction processes; generally, less flexible in terms of product types and adaptability to grain differences.

**short-time dough** — a straight dough to which a higher than normal proportion of yeast has been added to make the dough ferment faster.

**shred** — appearance of the surface within the break area. It is described as smooth, ragged, broken, etc.

**shredded wheat** — a ready-to-eat breakfast cereal made by pressing moist-cooked wheat kernels between corrugated rollers, collecting the shreds in a multitude of layers, separating the bed of layers into biscuits by a dull blade, then drying and toasting the biscuit.

## GLOSSARY OF CEREAL TECHNOLOGY TERMS    179

**Shrewsbury cakes** — *Old English* thin cookies made of a very rich unleavened dough, traditionally flavored with caraway seeds.
**shrink film** — a plastic film that can be draped around a package or product and then shrunk by applying heat, so as to get a tight fit.
**shrink tunnel** — the equipment used in the heat-treating stage of the shrink film packaging process.
**shui dian fen** — *China* a paste of cornstarch dispersed in water that is widely used as a thickener in sauces. Prior to the availability of cornstarch, arrowroot, tapioca, etc. were used.
**sieve** — a device for separating particles according to size; consisting of a frame or vessel having near its bottom a screen with uniform openings. zontal (rather than cylindrical) sieves are used.
**sifter** — an automatic sieving device, and particularly one in which horizontal (rather than cylindrical) sieves are used.
**sigma-arm mixers** — horizontal mixers with two arms of roughly rectangular cross-section that have been shaped into angled curves reminiscent of the Greek letter sigma.
**sigtebrod** — *Denmark* sourdough bread made with a blend of rye flour and wheat flour.
**sika** — *Greece* figs.
**silicone-coated paper** — paper that has received a coating of silicone antistick compound, often used as a pan liner.
**silicones** — compounds containing the element silicon, one type of which forms the active material in most of the liquids applied (often by spraying) on surfaces (such as the inside of bread pans) to reduce sticking of dough.
**silks** — the silky styles on an ear of maize; they are of little or no importance in dry milling or wet milling of field corn, but are regarded as blemishes or defects in preparations of sweet corn.
**silo** — a very large storage bin, usually assuming the form of a cylinder on end; in some cases, other shapes are used. Can be used for both dry and wet ingredients.
**silvano** — *Italy* a chocolate meringue piece or tart.
**silvapannukakku** — *Finland* pancake with diced bacon.
**silver pin noodles** — *China* a dough made of wheat starch mixed with boiling water is formed into nail-shaped noodles sbout 2 inches long by rubbing small balls of the dough across an oiled board.
**simighdhali** — *Greece* farina.
**simmer** — to cook a liquid so that just a few bubbles constantly form in the lower part of the vessel, and vigorous boiling is avoided.
**simnel cake** — a special cake formerly made for the Easter season; it resembles an English meat pie in shape but its composition is more like that of a rich plum pudding. Usually has a strip of marzipan or almond paste on the top.

**Simplesse** — trademark for a reduced calorie fat substitute useful as an ingredient in ice cream and the like, but not very useful in baked products.

**simple syrup** — two parts sugar mixed with one part water, heated and stirred until all the sugar has dissolved. A stock intermediate having a number of uses in the confectionery, baking, and food service industries.

**singin' hinny** — also, "singing honey." *UK* A rich sweet roll, containing currants, cooked on a griddle, then halved and buttered.

**single-acting baking powder** — a leavening system that generates carbon dioxide gas continuously when water is added to the dough or batter containing it; if not made up and baked promptly, mixtures containing this type of baking powder may exhibit insufficient rise.

**single lap oven** — a traveling tray oven in which the baking shelves travel one long horizontal run during baking.

**sinn** — *Egypt* flatbread made from a dough consisting mostly of bran. It is yeast-leavened and given one or more fermentation steps.

**sirap** — *Sweden* molasses and some other kinds of sweet syrups.

**siro** — *Mexico* sirup.

**siropi** — *Greece* syrup.

**six-rowed barley** — a species of barley, *Hordeum vulgare*, in which three kernels develop at each rachis node; it differs slightly from two-rowed barley in malting response.

**sizing** — in milling terminology, means breaking down and grading the coarser middlings.

**skaltsunakia sifneika** — *Greece* a turnover filled with fruit, chopped nuts, and cinnamon.

**skeins** — packaging units of spaghetti or vermicelli formed by twisting many strands of the long, moist pasta into figure-eight configurations before drying.

**skeppsskorpor** — *Sweden* hardtack.

**skim milk** — milk from which nearly all the butterfat has been removed by centrifugal separators.

**skinning conveyor** — a section of the processing line for, typically, marshmallow depositing, that allows the surface of the deposit to lose some moisture and thus become more receptive to a chocolate coating or the like.

**skive** — *Denmark* slice.

**skorpor** — *Sweden* biscuits. *Denmark* rusks.

**slabs** — marble or slate table tops used by confectioners as a surface on that to work boiled sugar and other hot mixtures.

**slack dough** — a dough having insufficient elasticity to process properly.

**slagroom** — *Netherlands* whipped cream.

**slankekost** — *Denmark* low calorie foods.

**slicers** — devices for cutting baked products, such as slicers for bread loaves; many types and sizes are available.

**slick test** — pressing and smoothing out a sample of flour and comparing its surface appearance to that of a standard sample, so as to qualitatively estimate the color, speckiness, etc. After the dry surface has been viewed, the sample is sometimes dipped in water to bring out additional details.

**slim cake** — *Ireland* a plain (i.e., lean dough, not garnished) cake.

**slow dough** — a dough that ferments and conditions slowly, usually due to a temperature that is too low, too much salt, insufficient or stale yeast, or too much sugar.

**slump** — a rather uncommon dessert resembling a fruit cobbler; it consists of a sweetened fruit (e.g., gooseberries, strawberries) mixture on which islands of biscuit dough have been deposited. This combination is baked and, when the biscuits are cooked through, the pan is inverted over a serving dish leaving the biscuits on the bottom.

**slurry** — a fairly concentrated aqueous mixture, but one sufficiently fluid to be handled by liquid systems.

**slurry mixers** — mixers for preparing slurries as a step in the forming of more complex blends; often with high speed agitators for putting poorly soluble powders into aqueous suspension before they are added to batter- or dough-mixers.

**småkage** — *Denmark* cookie.

**småkaka** — *Sweden* fancy cookie.

**smart valves** — valves for restricting fluid flow that exhibit some sort of internal or external automatic response to occurrences within the valve's environment.

**smead** — see "smid."

**smid** — *ME* finely ground semolina, often used as the basis for cakes and filled cookies in Middle Eastern cookery.

**smoke point** — the lowest temperature at which a fat sample, heated under a prescribed set of conditions, gives off a thin continuous stream of smoke. The smoke point of a frying fat should be as high as possible. It is typically dependent on the amount of free fatty acids and monoglycerides present in the oil.

**smör** — *Sweden Denmark* butter.

**smörbakelse** — *Sweden* pastry.

**smör bröt** — *Sweden* bread roll.

**smorkage** — *Denmark* coffee cake.

**smorodina** — *Russia* currant.

**smutter** — formerly, a general name for any wheat-scouring machine but now confined to a machine specifically designed to rid grain of smut, which is a type of microbiological growth.

**snack cake** — a portion-sized cake, usually iced and filled, intended to be consumed without eating utensils as a between-meals dessert, essentially filling the same role as candy bars.

**snack crackers** — crackers with savory flavors, often in unusual shapes, usually in sizes smaller than saltines. May be made up similarly to sandwich cookies, i.e., two crackers enclosing a filling such as peanut butter.

**snacks** — foods of almost any description that can be consumed without accompanying foods, in relatively small quantities, usually without table utensils, and which are not regarded as part of a complete meal.

**snaps** — small cookies that spread in the oven to give a thin layer which bakes into a crisp textured wafer.

**sneeuwbal** — *Netherlands* one kind of cream puff, sometimes filled with currants and raisins.

**soba** — *Japan* noodles made from buckwheat flour, usually with an admixture of wheat flour.

**socker** — *Sweden* sugar.

**socker kaka** — *Sweden* sponge cake.

**sock mai jai** — *China* baby corn, q.v.

**soda** — baking soda is sodium bicarbonate; washing soda is sodium carbonate.

**soda cracker** — also, "saltine." A thin crisp cracker made from a yeast-leavened dough with a late addition of baking soda; the dough is lean in formulation, has been laminated just before cutting, and contains a large amount of a sponge that has been fermented up to 16 hr.

**sodium acid pyrophosphate** — a common leavening acid that causes a relatively slow carbon dioxide release when combined with bicarbonate in a dough or batter.

**sodium aluminum phosphate** — a common leavening acid; slow acting.

**sodium aluminum sulfate** — one of the older leavening acids; slowest acting of all the common ones.

**sodium benzoate** — a preservative effective against many microorganisms when used in foods having a pH below about 4.5.

**sodium bicarbonate** — baking soda, $NaHCO_3$.

**sodium bisulfite** — also, "sodium metabisulfite." Functions as a dough conditioner to weaken or soften dough; in the past, it was added to most cracker doughs, but now it is out of favor, especially in the US.

**sodium diacetate** — a preservative described as a "molecular combination of sodium acetate and acetic acid." Its preservative effect is probably due to the release of free acetic acid.

**sodium erythorbate** — an antioxidant compound.

**sodium glutamate** — a flavor enhancer with very little flavor of its own.

**sodium propionate** — a preservative added to foods to delay mold growth; it has basically the same effect as propionic acid but is easier to disperse.

**sodium steroyl lactylate** — an emulsifier and dough conditioner with significant effects on processing response and finished product quality of many yeast-leavened and chemically leavened products.

GLOSSARY OF CEREAL TECHNOLOGY TERMS        183

**soft ball stage** — a stage in sugar boiling reached at about 245°F.
**soft meringue** — meringue with a relatively low proportion of sugar, used especially as thick layers on top of meringue pies; typically browned slightly on the top, but never baked.
**soft red winter wheat** — like other soft wheats, the flour from SRW wheat usually contains less protein and a lower quality protein than flour made from hard wheat. The flours from soft wheat do, however, cover a wide range of quality, and some are even used in family or all-purpose flour. Generally, SRWW flours are best for cakes and other pastries, cookies, snacks, etc., not so good for bread.
**soft water** — water with a comparatively low content of minerals, especially calcium. Distilled water is the ultimate in "soft" water.
**soft wheat** — those cultivars of *Triticum aestivum* normally yielding kernels that, compared to hard wheat, are easier to crush and have a whiter endosperm breaking into smaller particles.
**soft wheat flour** — the flour milled from soft wheat, often whiter, finer in particle size, and weaker than flour from hard wheat.
**sokeri** — *Finland* sugar.
**sokerikakku** — *Finland* sponge cake.
**sokolata** — *Greece* chocolate.
**sol** — *Russia Serbia/Croatia* salt.
**solid fat content** — also "SFC." The actual percent of solid fat in a sample of fat or oil, determined at standard temperature check points by pulsed nuclear magnetic resonance procedures. This analytical method is based on the assumption that protons in liquid fat are more mobile than those in solid fat. These more mobile protons are responsive to a magnetic field. Compare with solid fat index.
**solid fat index** — also, "SFI." An empirical measure of the solid fat content of a sample of oil conducted at standardized temperature check points. It is based on a dilatometric procedure relying on volumetric changes occurring during melting or crystallization. SFI specifications are important in determining the behavior of a fat during processing, etc.
**solid heat** — when all oven components have been brought to the temperature they will maintain during continuous production, they are at solid heat. The term means that the temperature can be maintained within the desired range even when product is going through the oven at the top rate.
**solomka** — *Russia* sugared or salted cookies or crackers, often in very large sizes.
**solubility index test** — an evaluation applied to dried milk products for the purpose of determining ease and extent of reconstitution.
**solution** — a liquid in which another substance has been completely dispersed in ionic or molecular form. If molecular aggregates exist, they are not dissolved.

**somen** — *Japan* thin noodles (vermicelli) made from wheat flour.
**somun** — *Serbia/Croatia* a round flat bread.
**sonho** — *Portugal* a kind of doughnut.
**sooji** — *India* semolina.
**soola** — *Estonia* salt.
**sopapillas** — *Spain* chemically leavened small deep-fried cakes, somewhat like a doughnut without the hole; usually only slightly sweet.
**sorbates** — see sorbic acid, of which sorbates are the salt form. Sorbates generally dissolve more readily than sorbic acid.
**sorbic acid** — a preservative effective against many molds, yeasts, and bacteria. Depresses activity of baker's yeast and so is not commonly used in yeast-leavened doughs. Formulators should check the flavor consequences of using it when they expect their product to be heated, especially toasted, by the consumer.
**sorbitan esters** — emulsifiers made by reacting sorbitol esters with fatty acids, such as sorbitan monostearate.
**sorbitol** — a sugar alcohol used as a humectant or water binder in confections and bakery foods.
**sorghum** — a cereal grain mostly used as animal feed, but also as a human food in some parts of the world. A different variety of the plant is used as a source of juice that is evaporated to form sorghum syrup.
**sorghum sugar** — a solid sweetener made from the juice of sweet sorghum stalks, not a common item of commerce in the US.
**sorghum syrup** — a specialty table syrup and sweetening ingredient still being used to some extent in the Southern US; it is made from the cane juice of sweet sorghum that has been condensed by boiling. Sucrose is the main constituent.
**sour cream** — liquid dairy product of high butterfat content that has been cultured with bacteria to develop an acidic taste and a thick texture.
**sourdough** — (1) A starter or leaven containing some basic dough ingredients plus a large population of yeasts and bacteria. Doughs made with starters typically develop a sour flavor and pungent aroma. Many, perhaps most, sourdoughs contain large amounts of lactic acid bacteria. (2) As an adjective applied to bakery foods, "sourdough" generally identifies products that have, at some point in their preparation, been subjected to a fermentation long enough to allow the generation of lactic acid and other flavoring materials thought to be characteristic of such products.
**sours** — commercially available ingredients usually containing lactic and acetic acids as well as other flavoring materials that have been developed by yeast and bacteria growing over a period of time; used as flavoring and/or leavening ingredients, particularly in rye breads.
**soybeans** — a legume (not a cereal) that produces medium-size round seeds containing fairly high amounts of protein and oil. The beans are generally

## GLOSSARY OF CEREAL TECHNOLOGY TERMS 185

pressed and extracted to remove the oil, and the cake that remains is converted to animal feed or further processed to yield protein supplements and other food ingredients.

**spaetzle** — also, "spatzli" and "spatzle." *Germany* A type of noodle usually seen as irregular lumps or short strands; often formed from a fairly stiff egg dough by manually forcing it through holes in a colander or similar utensil; spaetzle are typically cooked in soup much like noodles.

**spaghetti** — thin, round smooth strings of pasta; can vary in diameter and length, but never have central holes or external ridges (other names are used for the latter forms).

**Span** — a trade name for sorbitan monostearate, an emulsifier that has found some applications in chemically leavened foods.

**spanakopita** — also, "spanikopita." *Greece* A casserole dish of spinach, feta cheese, eggs, and onions wrapped in (or layered with) phyllo and baked.

**spatula** — a utensil shaped like a knife with (usually) a rounded end, but not sharpened; flat, thin, somewhat flexible, and often with a plastic or wooden handle.

**specific gravity** — the ratio of the weight of a given volume of sample material at 25°C to the weight of the same volume of water at the same temperature.

**specific heat** — the ratio of the quantity of heat required to raise the temperature of a body one degree to that required to raise an equal mass of water one degree.

**specific volume** — the reciprocal of specific gravity. An important factor in bread quality evaluations, where the figure is ordinarily expressed as the number of cubic cm occupied by one gram of substance.

**speculaas** — *Netherlands* spiced almond cookie.

**spekulatius** — a butter cookie of a type common in Germany but made in many other countries; it is characteristically molded in wooden forms, usually appearing as rectangles with human or animal figures on top. Commonly flavored with spices.

**spelt** — a primitive type of wheat, of no commercial importance.

**spettekaka** — *Sweden* cone-shaped cake made by baking dough or batter deposited on a spit rotating in front of a heat source.

**spice** — any one of large number of materials derived from plants (seeds, flowers, etc.) that contain odoriferous constituents suitable for use as food flavorings.

**spiedini alla Romano** — *Italy* alternating cubes of bread and cheese threaded on a skewer and baked.

**spindle mixer** — mixers used mostly for cracker doughs that include two or more vertical shafts bearing auger-shaped mixing blades that can be lowered into specially configured dough troughs.

**spiral cooler** — equipment for chilling or freezing foods that carries the product on a continuous conveyor forming a helical pattern about a central cooling duct.

**spiral mixer** — mixer with an agitator having an approximately spiral configuration, usually working in a wide, fairly shallow bowl..

**sponge-and-dough** — a method of producing bread in which a sponge is made by mixing a part of the flour with part or all of the water, and, usually, all of the yeast and yeast food; this mixture is allowed to ferment until it is judged to be ready for incorporation with the other ingredients to make the "dough."

**sponge cake** — a cake without shortening, or with very little shortening, made by combining a whipped mixture of whole eggs or egg yolks with the other ingredients, with or without added chemical leaveners.

**sponge fingers** — *UK* ladyfingers.

**sponge flour** — a flour for use in preparing bread sponges; may be either the same kind of flour as used for the dough or a stronger flour.

**spoon bread** — a type of corn bread cooked in a casserole that has a high moisture content so it exhibits a very soft consistency, in some cases almost like a pudding or a souffle. It is sometimes highly seasoned, and may include particles such as corn kernels, chopped sweet peppers, etc.

**spore** — one of the reproductive forms of certain fungi, bacteria, and yeast; microscopic, dormant, often spheroidal forms that germinate when provided with suitable growing conditions. They are much harder to destroy than the vegetative form of the same microorganism. Mold spores are found everywhere in the environment.

**spotted dog** — also, "spotted dick." *UK* a cylindrical steamed pudding composed mostly of suet and flour and containing raisins or currants.

**spotty heat** — the condition existing when an oven has a very uneven heat distribution.

**spray drying** — the process of atomizing a fluid food product and delivering the mist to a chamber through which hot air is flowing. The resultant low-moisture powder has minimum heat damage and is usually in the form of a fine powder. Has been applied to many foods, such as eggs, milk, yeast, and cheese, to form storage-stable bakery ingredients.

**sprayed cracker** — the type of savory cracker (rich saltine type) that is sprayed with vegetable oil after it is baked so as to improve texture and appearance; often flavor is also improved by this treatment.

**spread** — an important specification applied to cookie flours; it is based on the change in width of the test dough piece as it bakes.

**spring** — see, "oven spring."

**springerle** — Bavarian-type anise-flavored sugar cookies, formed by pressing dough into wooden molds, then air-drying the pieces before baking.

**spring rolls** — also, "egg rolls." Thin, barrel-shaped pastry packages of a few ounces in weight stuffed with savory fillings and deep-fried; the wrappers vary from thin egg-based crepes and wafer-thin rice flour discs to the much more common (in the US) commercially prepared flour dough sheets. A specialty of Chinese cuisine.

**spring wheat** — wheat (*Triticum aestivum*) that is sowed in the spring and harvested in the fall; generally has higher protein content and yields a stronger flour than winter wheats.

**sprits** — *Netherlands* spritz cookies, usually made of an air-leavened or unleavened dough, sometimes of the shortbread type.

**spritz** — (1) A method of forming cookie doughs by extruding a fairly soft batter through a tube so that some surface detail is retained from the orifice outline; also applied to similar extrusion processes used for other food materials. (2) The cookies so made, often seen in fanciful shapes.

**sprouted wheat** — fully matured wheat kernels that have been wetted while still on the plant, and have sprouted before being harvested. The baking quality of flour milled from such wheat is very bad. Some health food enthusiasts deliberately sprout wheat grain to make a bakery ingredient thought to have some dietary or spiritual advantages.

**spun sugar** — sugar syrup that has been cooked to about 300°F and spun into threads by a mechanical process or by dipping a bundle of wires into the molten mass and waving it in the air. Used only for decorations. Cotton candy is made by a variation of this process.

**squalene** — an isoprenoid that is a precursor of sterols such as cholesterol; found in the oils of some grains and in olive oil and fish liver oils.

**squashed spiders** — *UK* a sort of large cookie made of relatively lean shortbread dough or pie dough that has been mixed with a type of mincemeat before baking.

**stability** — the relative resistance of a product to undesirable change. For fats and oils, stability may refer to resistance to such deterioration as oxidation, hydrolysis, flavor reversion, or development of off-flavors and off-odors. For doughs, the term generally refers to the range of fermentation time through which an acceptable product can be obtained. For many food products, such as bread, stability includes (among other factors) the resistance to deleterious microbiological processes.

**stabilizer** — any additive causing a food product to retain a desired characteristic for a longer period of time, as by delaying the collapse of foams or the breaking of emulsions. Bread stabilizers may have the function of delaying development of undesirable firmness (texture staling).

**stack ovens** — deck ovens that can be stacked one on top of another so as to save space.

**stafidha** — *Greece* raisins.

**staksill** — *Sweden* semolina.

**stale** — a finished food product that has undergone physical and chemical changes rendering the product unacceptable from an organoleptic standpoint. Does not necessarily mean the product is harmful or lacks nutritional value.

**staling** — decrease in acceptability due to changes occurring after production; the changes may be due to chemical reactions or physical modifications, but "staling" is not generally regarded as including obvious mold growth (spoilage) or mechanical damage due to handling abuse (breakage).

**stamping machines** — devices for cutting dough sheets into shaped pieces by means of a reciprocating (up and down movement) die plate, sometimes incorporating a separate docker plate. Much used in cutting soda cracker doughs and the like.

**Standard Deviation** — a statistical term indicating the square root of the mean square of the deviations from the mean of a population of samples. Also called "sigma."

**starch** — a polymer of glucose found in many kinds of plants. Digestible by humans and forms a large part of the caloric content of many foods. When offered as an item of commerce, the ingredient generally is in the form of a fine white powder, dispersible but not soluble in water.

**starch acetate** — starch molecules that have been chemically acetylated to form thickening ingredients.

**starch conversion** — the chemical processes used to change starch granules into sweeteners such as corn syrup and non-sweet products such as maltodextrin.

**starch damage** — physical changes (breakage, abrasion) of starch granules, usually an inadvertent result of milling or some other process. Such granules are more susceptible to attack by amylases, and the amount of starch damage in a flour has been shown to be a factor affecting dough and bread quality.

**starköl** — *Sweden* beer with high alcoholic content.

**steak and kidney pie** — an English main-course pastry consisting of an crust (usually unleavened) containing a large amount of gelatinous gravy and a small amount of the named ingredients.

**steam** — vapor emitted from water at its boiling point; confined water vapor held at a temperature above the boiling point at the pressure existing in the enclosure.

**steamed bread** — *China* very similar to large dumplings in form and general appearance; dough lumps that have been cooked by hot water vapor.

**steaming** — (1) Injecting steam into an oven while products are being baked. (2) Cooking products by holding them in an atmosphere of hot water vapor — applied to dough products such as certain kinds of puddings, brown bread, etc.

**steam injection** — the operation of piping steam into oven chambers to secure special crust characteristics on bread and rolls.

**steam kettle** — versatile cooking equipment consisting of an inner vessel

## GLOSSARY OF CEREAL TECHNOLOGY TERMS 189

surrounded by a jacket (or pipes) through which steam is circulated. This equipment is made in a wide range of sizes and is found in many food plants including bakeries.

**steam tables** — omnipresent in cafeterias and most other restaurants, but also used in bakeries to keep icing, etc., warm. They consist of a table with openings into which containers of food can be put; below the top deck is a chamber containing water heated by some means, e.g., by low pressure steam, to keep a more-or-less constant (below boiling) temperature.

**steam-tube ovens** — ovens in which the baking chamber is heated by pipes containing steam under high pressure.

**stearine** — generally, a hard fat. Can be an oil that has been highly hydrogenated or the high-melting fraction from a rendering process, as in beef tallow rendering, where the other fraction is oleo oil.

**steep** — to put some solid substance in water for an extended period of time (hours) so as to cause the particles to become thoroughly equilibrated with the solution, as in steeping barley grain in the preparation of malt.

**steepwater** — the liquid drawn off from steeping vats; it contains soluble substances extracted from the grain as well as insoluble materials; it is generally either discarded or dried to make an animal feed ingredient.

**stellette** — *Italy* pasta in small star-shaped pieces.

**sterilize** — reduce the viable microbial content of a material (or space) to essentially zero by heating, irradiating, adding chemicals, etc.

**stevioside** — a glycoside extracted from the leaves and twigs of a plant found in South America, about 300 times as sweet as sucrose.

**stick pretzels** — thin, fairly short rods of (generally) yeast-leavened dough that have been dipped in an alkaline solution, sprinkled with coarse salt, and baked almost to dryness.

**sticky doughs** — doughs that have excessive surface adhesion, a quality that makes them difficult to process. Excess water in the formula is a common cause, but too much punishment in the make-up equipment will have a similar effect, as will excessive mixing, under- or over-fermentation, hot doughs, and several other processing or formulation errors.

**stiff dough** — a tough, rubbery, bucky bread dough.

**stollen** — a moderately flat multi-serving loaf made from a rich yeast-leavened dough containing pieces of candied fruit, raisins, nuts, etc.; it is usually lightly flavored with spices. Often iced or frosted after baking.

**stone sheller** — a dehulling machine for rice kernels, generally in the form of two horizontally-aligned abrasive-coated disks separated by about the length of a kernel; the upper stone is stationary, while the lower is rotated.

**storage life** — the time between completion of processing and the development of some unacceptable quality factors, i.e., staleness.

**storage proteins** — proteins that are more or less stable in the living cell, that is they do not participate in the cycles responsible for growth and reproduction of the plant cell; gluten is a storage protein complex of wheat.

**storage temperature** — the temperature (usually average temperature, sometimes a range) at which a product or ingredient has been, or should be, stored.

**stout** — *UK* a dark, sweet, viscous, beer of relatively high alcoholic content.

**straight dough process** — a method of bread-making in which substantially all the ingredients are mixed together at one time and then fermented, as contrasted with the sponge-and-dough process.

**straight flour** — all of the mill grind that can be designated as flour, what remains is millfeed. It can be divided into patents and clears, q.v.

**strain-gauges** — small-scale circuits designed to change in electrical properties as their shape is modified by outside forces; used as the sensing elements in weighing devices based on load cells.

**straps** — two or more loaf pans bound together by metal "straps" so they can be handled as one unit in the breadmaking process.

**stratify** — to spontaneously separate into layers, as when emulsions break down into an aqueous layer, an oily layer, and a solid layer. Making layers of dough intentionally is, however, called "laminating."

**strength** — a rather inexact term applied to the baking quality of flour; it summarizes the contribution of the flour to extensibility, absorption, stability, and other characteristics of the dough as well as the specific volume and texture of the baked product. Strength depends to a large extent upon the percentage of gluten in the flour, although other factors are involved.

**strengthener** — a wheat that is added to another wheat in a mill mix, to improve the quality or quantity of its protein content.

**streusel** — a crumb-like mixture of indefinite composition but almost always containing at least sugar, flour, and fat. Used for strewing on the top of the fruit in pies and coffee cakes, on crumb cakes, etc., usually before baking. Other kinds of mixtures have also been called streusel.

**streuselkuchen** — *Germany* coffee cake covered with streusel.

**string icer** — a machine that deposits thin strings or strips of warm icing on the top of sweet rolls, coffee cakes, etc. The more advanced equipment can automatically move the nozzles to give a simple design.

**stroop** — *Netherlands* sweetener syrup, molasses.

**strucla** — *Poland* yeast-leavened sweet rolls, often with raisins or fruit or nut pieces, etc.

**strucla owocowa** — *Poland* fruit rolls, dough rectangles spread with jam jelly roll fashion, but before baking.

**strucla z makiem** — also, "strucla z makowiec" or "strucla z makownik." Poppyseed rolls, often including bits of dried fruits and the like.

**strudel** — a product prepared by rolling paper-thin flour dough (often many layers) into cylinders containing fruit or other fillings, then baking.

**strudel leaves** — squares of unbaked very thin flour and water dough;

some versions include a small proportion of whole eggs.

**strudla** — *Serbia/Croatia* strudel.

**stuffed straight** — a flour consisting of a straight flour to which has been added clears from another milling operation.

**stuffing mixes** — blends of bread crumbs or cubes, spices and other seasonings, and other ingredients, intended for use by the consumer as a stuffing for poultry or the like after mixing with water or broth.

**sublimation** — emitting of vapor from a frozen (i.e., solid, not liquid) material; the way water is lost from ice in frozen products.

**submerger-plate fryers** — continuous fryers, such as used for doughnuts, in which the conveyor has means for keeping the buoyant dough piece below the fat surface for at least part of the cooking time.

**succinylated monoglycerides** — surfactants sometimes used in doughs to improve handling properties, soften crumb texture, and increase shelf life.

**sucralose** — a synthetic bulk sweetener or sugar replacer that has low or no caloric content.

**sucre** — *France* sugar.

**sucre en poudre** — *France* a fine granulation of sugar.

**sucre glace** — *France* sucrose of very fine particle size, approximately equivalent to confectioners', powdered, and 10X sugar.

**sucre semoule** — *France* sucrose in large particles, often used in making fruit preserves.

**sucrose** — the chemical substance of which table sugar, beet or cane sugar is about a 99% pure example. When hydrolyzed by invertase or mild acid treatment, sucrose yields the hexoses, fructose and glucose.

**sucrose esters** — sucrose molecules combined with 1, 2, or 3 fatty acids. These compounds have been recommended as dough strengtheners and crumb softeners.

**sucrose polyester** — a fat-replacer synthetic having reasonable stability at high temperatures, thus of possible use as a low/no calorie frying fat.

**suet** — mostly *UK*. Beef fat prepared for, and used as, a shortening in puddings and bakery foods; the cattle counterpart of lard. This term is seldom encountered in the U.S., though the same material is used in shortenings and frying fats.

**suet pudding** — *UK* a boiled or steamed pudding made of chopped suet, flour, bread crumbs, sugar, and milk; often containing raisins or other fruit; may be served with a sweet topping (jam) to be added by the diner.

**sugar** — when not modified, means a substantially pure preparation of cane or beet sucrose. In the most general sense, means any carbohydrate having a sweet taste. The sap of various trees, certain type of maples and palms, for example, is also used as a source of sugar throughout the world.

**sugar alcohols** — compounds formed by hydrogenating sugars; examples are sorbitol, mannitol, and xylitol. They may be sweet but are generally nonfermentable; they have various food uses.

**sugar batter procedure** — a method of mixing cake batters that requires creaming the shortening with sugar, then blending in the eggs, colors, and flavors, and finally mixing this with flour and leaveners.

**sugar crisps** — a confection, one version being a thin, fairly large disc of sweet dough or puff pastry that has been heavily coated on both sides with granulated sugar and cinnamon before being baked almost to dryness.

**sugar paste filling** — adjuncts consisting mostly of powdered sugar and fat or oil, that are plastic when hot but firm up as they cool.

**sugar rest** — in brewing, the period during which the mash is held for about 5 to 30 minutes at 133°F to 144°F. a temperature near the optimum for beta-amylase.

**sugar wafers** — cookies and cookie bases made by baking unleavened batters of very low viscosity between two heated plates bearing a design, then interleaving two or more of these bases with a "creme" consisting mainly of a fatty material and finely ground sugar. Both the separate baked product and the finished creme cookie have been called "sugar wafers."

**suhkrut** — *Estonia* sugar.

**suiker** — *Netherlands* sugar.

**suizo** — *Spain* bun, roll.

**suji** — *India* semolina.

**suji beveca** — *India* a sweet dessert made from semolina.

**sukhar** — *Russia* rusks.

**sukker** — *Denmark* sugar.

**suklaa** — *Finland* chocolate.

**sulfhydryl groups** — -SH groups that, on protein molecules, are responsible in part for the changes in physical properties of doughs observed when oxidation or reduction occurs.

**sulfite** — a reducing agent, often in the form of sodium metabisulfite, that has been used to "relax" stiff doughs.

**sulfur dioxide** — a gas that has been used as a bleaching agent by some food processors such as manufacturers of dried apricots.

**sultanas** — a type of raisin, common term in the U.K., not in U.S.

**summer pudding** — *UK* a dessert made from a crust or casing of white bread slices, into which is placed sweetened cooked fruit. The combination is not cooked.

**sunflower oil** — oil that has been pressed or solvent-extracted from seeds of the sunflower. Contains around 69% polyunsaturated fats.

**suola** — *Finland* salt.

**superheated steam** — steam that does not contain as much water as could be held in a similar volume at equal temperature and pressure.

**surbrod** — *Denmark* sourdough bread, of several types.

**surfactants** — surface active agents, i.e., chemical substances capable of reducing the surface tension of water. Important for their ability to improve emulsification of water and oil. Surfactants have been added to doughs, batters, and adjuncts to improve their processability and increase shelf-life of the finished products.

**sur levain** — *France* a sourdough method for making French bread.

**sur poolish** — *France* a sponge method for making French bread.

**suspiro** — *Portugal* meringue; meringue cookies.

**sussigkeit** — *Germany* a candy or confection.

**su tang** — *China* long thin roll of gluten dough, somewhat resembling a croissant in shape. The consumer slices this into discs which are cooked in stir-fried and braised dishes as a meat substitute.

**suvi kolaci** — *Serbia/Croatia* cookies.

**sveske** — *Denmark* prune.

**sveskebrod** — *Denmark* yeast-leavened dough containing an appreciable amount of chopped prunes; usually made with white flour.

**svingi** — *Greece* fritters.

**sweating** — the changes in wheat induced by moisture and heat, as in tempering or during storage. Principally due to biological changes such as respiration, but also may include chemical and physical processes, also spoilage processes mediated by contaminating microorganisms.

**Swedish crispbread** — thin, crisp, dark wafers made from rye and/or whole wheat flours; in modern plants versions, made by extrusion and not yeast-leavened.

**sweet chocolate** — an eating and ingredient chocolate containing at least 15% chocolate liquor plus sugar and flavoring materials but no dairy products; must conform to Federal Standards of Identity.

**sweet corn** — varieties of maize that have tender kernels relatively high in sugars, suitable for eating as a vegetable on or off the cob.

**sweet doughs** — enriched yeast-leavened doughs containing relatively high contents of sweeteners and, usually, shortenings; may also include ingredients selected from very many other modifying and improving agents, decorative additives, flavors, colors, etc. Used as a basis for all sorts of coffee cakes, sweet rolls, and the like. The term is generally not applied to true Danish pastry doughs, puff pastry doughs, or doughnut doughs, although this is by no means an inflexible practice. Less freqently applied to chemically leavened doughs used for the same purposes.

**sweetened condensed milk** — whole milk mixed with sugar and reduced in moisture content by heat evaporation. There is also a skim milk version. Federal standards apply. Has been used as an ingredient for confections.

**sweeteners, non-nutritive** — sugar substitutes such as saccharin that do not contribute any nutrititive value (including calories) to humans.

**sweetmeats** an old and very general term for confections, candies, glacé fruits, etc.

**sweet paste** — a modification of shorbread cookie dough that includes a small amount of whipped eggs to give a slight expansion during baking.

**sweet sorghum** — or, "sorgo." A type of sorghum, the stalks of which contain sweet juice at some point during its growth and are suitable for making sorghum syrup.

**sweet whey** — whey derived from a cheese-making operation that does not yield a reaction mix of moderately high acidity. The preferred form of whey for bakery products.

**Swiss roll** — *UK* a jelly-roll type of confection consisting of a layer of baked cake coated with jam, jelly, marshmallow, or other filling, then rolled up to give a spiral cross-section; usually enrobed with chocolate couverture.

**sylt** — *Sweden* jam.

**syltoj** — *Denmark* jam.

**symmetry** — the extent to which a loaf is uniformly proportioned.

**syneresis** — the gradual separation of gels into a liquid portion and a shrunken solid (plastic) portion. Happens to virtually all food gels, given sufficient time.

**synergist** — a substance that, when used in combination with another material, results in a combination having a greater effect (of some sort) than would be predicted from the sum of the effects of the individual materials.

**synthetic** — a substance made artificially by combining (usually by chemical reactions) other materials; not appropriately applied to simple mixtures, however.

**syrups, blended** — ingredients consisting of sugar syrups and corn syrups mixed in some proportion, usually for economic reasons.

**szarlotka** — *Poland* an apple cake.

## -T-

**taart** — *Netherlands* cake.
**table syrup** — flavored sweeteners in the form of a thick liquid, intended to be applied to pancakes and the like by the ultimate consumer.
**taco** — *Mexico* a tortilla wrapped around some sort of savory filling; dozens if not hundreds of major variations exist, such as taquitos (smaller diameter), flautas (still smaller), enchiladas (covered with sauce), etc.
**taco shell** — a tortilla, usually of medium- to large-size, fried (usuually after it has been folded into a U-shape) until it is crisp.
**taffy** — a plastic, sticky confection usually made of a sweetener syrup that has been boiled to a low moisture content, then repeatedly folded and stretched so as to incorporate some air. Ingredients may include some shortening and flavors. Many variations are extant.
**tærte** — *Denmark* cake, tart.
**taftoon** — *Iran* round, sourdough flat bread with small holes on its surface.
**tagliatelle** — *Italy* flat noodles.
**tagliolini** — *Italy* thin, flat noodles.
**tahini** — *Middle East* untoasted white sesame seeds, ground to a paste.
**tailings** — material that is too coarse to pass through the screens in a bolter or sifter.
**tail of the mill** — the part of the mill where the final processing occurs.
**tail of the mill particles** — wheat particles of relatively small size that are removed near the end of the milling plant.
**tail sheet** — a coarser screen at the far end of a sifter, intended to scalp stock that is coarser than the material screened out by the rest of the sieves.
**tainas** — *Estonia* dough.
**tallow** — hard beef fat obtained by rendering fatty tissues removed during butchering; sometimes also applied to fat from sheep, etc. There are edible and inedible forms, the latter mostly used as a raw material for soap. Tallow contains a mixture of oils and hard fats of potential value as shortening ingredients, e.g., oleo oil and stearine.
**talmouse** — *France* cheesecake.
**tamale** — masa spread in a fairly thin layer on a corn husk, partially covered with comminuted or shredded meat and sauce, then rolled up to make a cylinder having meat in the center. The husk holds the assemblage together during further cooking (usually by boiling) and is removed by knowledgeable consumers before the tamale is eaten.
**tâmara** — *Portugal* date.
**tamarind** — flat, dark brown, bean-like pods collected from the tree *Tamarindus indica*. Intensely bitter and sour. Used as flavorings in many

sauces in Asian cuisines. Also, as the basis of a confection of limited appeal made by coating the pods with powdered chilis and sugar.

**tamatar** — *India* tomato.

**tamees** — *Saudi Arabia* thin flatbread made from a very lean yeast-leavened dough.

**tandoor** — also, other spellings. *India* a barrel-shaped oven made of mud or clay, used for baking bread, roasting meats, etc.

**tandouri roti** — *Pakistan* a flat bread, the preparation method and formula for which are similar to those used for chapati, but baked in a tandoori (tanduri), an oval in-ground oven having its walls plastered with clay.

**tannin** — polyphenols found in plant tissues, often bitter in taste and brown in color; may affect the flavor of certain sorghum grain, etc. It is also important component of hops.

**tannouri** — *Saudi Arabia* round flatbread made from a lean yeast-leavened dough that undergoes one fermentation and one proofing stage. Has a dark-spotted, golden brown crust, little or no crumb, and a docked upper surface.

**tanok** — *Iran* round flatbreads baked in the countryside from water, salt, flour, and starter; they vary in thickness from 3.5 cm to paper thin.

**tapa** — *Spain* snack or appetizer.

**tapioca** — a food material (largely starch) obtained from roots of cassava plants, *Manihot utilissima*. At one time, it was widely used to prepare puddings in home kitchens. Tapioca pearls and chips are still widely used in Asia as ingredients in puddings, cookies, and confectionery. The raw material can be further processed to yield an industrial starch with some unusual properties.

**tapsi** — *Greece* a metal sheet on which cookies, tarts, etc. can be baked.

**tapsi ghlika tu tapsiu** — *Greece* describes any pastry baked on a tapsi.

**tap water** — generally means potable water delivered by a municipal supplier through its piping system.

**taro** — a plant, *Colocasia esculenta*, the tuber of which is widely used as a food raw material throughout the Pacific islands and elsewhere, and is the source of a specialty starch.

**tarta** — *Spain* cake, tart.

**tärta** — *Sweden* cake.

**tartaric acid** — an acid obtained as a by-product of wine preparation, and from other sources. Used as a mild acidifier in food products; was at one time a common acid-reacting component in baking powders.

**tarte** — *France* open faced pie or tart.

**tarteletter** — *Denmark* patty shells.

**tarte Tatin** — *France* upside-down tart of caramelized apples.

**tarts** — small pies, usually sized for individual servings.

**tascheri** — *Germany* pastry turnover with meat, cheese, or jam filling.

**taste** — the sensations sweet, sour, salt, and bitter. Some authorities claim the term should include other mouth and tongue sensing capabilities, e.g., chemical heat sensations (as from pepper). Not properly used to include aromatic properties that are part of the flavor complex, however.

**täyte kakku** — *Finland* layer cake.

**TBHQ** — a synthetic chemical substance effective in retarding development of oxidative rancidity, similar in its action to BHA and BHT.

**tea cake** — *UK* sweet roll.

**tea rolls** — small sweet buns.

**tear strength** — a test applied under standardized conditions to give a measure of the resistance of packaging film to tearing after a hole or slit has been made in the film.

**tebirkes** — *Denmark* bun with poppy seeds.

**teff** — tufted annual grasses somewhat similar to sorghum and millet; *Eragrostis tef* or *Eragrostis abyssiniea*; important food grains in Ethiopia,

**Teflon** — tradename for a white waxy plastic that has low friction and very low adhesive properties; used as a coating or casing for preventing the sticking of doughs and the like to processing machine surfaces.

**teig** — *Germany* dough.

**teigwaren** — *Germany* macaroni products, alimentary pastes.

**tejolote** — *Mexico* a pestle used with a molcajete to grind nixtamal for tortillas; usually wide at bottom and narrow at top, like a typical lab pestle.

**tempe** — *Malaysia Indonesia* soybean cake highly modified through reactions mediated by a mixed microorganism inoculum.

**tempering** — similar in meaning to conditioning, but applied occasionally to materials other than grain. Also, can mean allowing an ingredient or product to come to the same temperature as the space in which it is stored.

**tempura** — *Japan* pieces of vegetables, meat, fish, etc., coated with a batter typically made of wheat flour, water, and egg yolks, then deep-fried.

**tempura ko** — *Japan* flour used to make the tempura batter for coating shrimp and raw vegetables that are to be deep-fried. Often, a low-gluten wheat flour.

**tenderness** — used by bakers to describe the fragility or relative mechanical strength of the crumb or crust of a baked product.

**tensile strength** — the greatest longitudinal stress a substance can bear without tearing apart, usually expressed with reference to a unit area of cross-section.

**teosinte** — a wild grass found in central and south America that is thought by some scientists to be an ancestor of maize.

**terabelesi** — *Tunisia* round flatbread made from a flour, water, salt, and yeast dough that is fermented and proofed. Cuts forming a square are made across the top of the dough piece before it is baked.

**terminal velocity** — rate of aspiration or air flow that is just sufficient to lift a given type of particle from a mixture, usually reported in ft per min.

**terveysruoka** — *Finland* health food.

**testo** — *Serbia/Croatia* dough.

**test weight** — for wheat, the weight in avoirdupois pounds of a Winchester bushel of the grain.

**tészták** — *Hungary* dumplings.

**texture** — the mouthfeel, consistency, or other effects a food product has on the tactile senses, including those involved in chewing, described by such words as elasticity, viscosity, toughness, and smoothness. When applied to bread crumb, however, "texture" usually means only the sensations obtained when the fingertips of the examiner are lightly passed across a freshly cut surface, these being typically described by terms such as smooth, silky, harsh, etc.

**texture staling** — the undesirable changes occurring upon aging of, e.g., a loaf of bread that are principally recognized as increased firmness and harshness of the crumb to the touch, and a reduced smoothness and elasticity in the mouth.

**thaumatin** — a very potent sweetening substance derived from the African katemfe fruit.

**thermal destructors** — in odor control, equipment that subjects vapors to temperatures high enough to convert most of the organic substances to carbon dioxide and water.

**thermoduric** — describes microorganisms that can withstand high temperatures.

**thermoplastic** — when applied to resins, indicates they respond by softening or even liquefying when heat is applied, a characteristic of most plastics that are used in making food packages.

**thermosetting** — when applied to plastics, identifies those reaction mixtures that become permanently rigid when sufficient heat is applied, thus opposed to "thermoplastic." Such plastics are employed primarily as materials for baking pans, machine construction, and the like.

**thermostat** — the combination of sensing and activating mechanisms used to maintain the temperature of a gas or liquid within a narrow range by modulating a heating and/or cooling source.

**thiamin** — a B vitamin, one of those that must be added to enriched flour.

**thinh** — *Vietnam* toasted, ground rice, used to provide texture and flavor to ground pork and the like.

**throughs** — particles small enough to pass through the openings in a bolter or sieve.

**tighanita** — *Greece* fritter.

**tighanito** — *Greece* fried.

**tight** — baker's term for a stiff, tough, or bucky dough.

**tikerberikook** — *Estonia* gooseberry cake.
**til** — *India* sesame seeds.
**tillage** — the improving or cultivation of land for agricultural purposes.
**tillering** — the forming of new shoots on stems of, e.g., wheat plants.
**timbale** — *France* a pastry case or mold cooked with a filling of meat, fruit, vegetables, etc.
**tin bread** — *UK* pan bread.
**tines** — (1) The prongs of a fork or the like. (2) Pointed metal rods used in equipment designed for fork-splitting English muffins.
**tinplate** — thin steel sheets covered on one or both sides with a very thin coating of tin, used for making food cans, baking pans, etc.
**tip caps** — fibrous structures, often dark brown or black, found at the pointed end of corn kernels; they are discarded in all modern forms of corn milling; they form an undesirable part of popped corn.
**tiramisu** — *Italy* a dessert consisting typically of ladyfingers soaked in coffee, covered with a mixture of (e.g.) mascarpone cheese, egg yolks, and sugar, and topped with cocoa powder or grated chocolate.
**tiropita** — *Greece* cheese tart.
**tiropitaki** — *Greece* small cheese tarts.
**titanium dioxide** — a non-certified color, insoluble in water, that is useful in whitening icings and the like.
**tocino de cielo** — *Spain* (1) Custard-filled cake. (2) Caramel mold.
**tocopherols** — any of a number of alcohols of high molecular weight having the properties of vitamin E in varying degrees; found in wheat germ, crude cottonseed oil, etc., also prepared synthetically.
**toffee** — a form of caramel candy, often cooked quite hard (i.e., to a low moisture content).
**tofu** — *Japan* soybean curds, typically formed into soft white briquettes with some of the moisture pressed out; in Asian cuisine, it is added to many soups and other dishes; has been suggested as an ingredient in health breads and as a nondairy replacement for yogurt or cheese.
**tomate** — *Spain* tomato.
**toost** — *Netherlands* toast.
**top crust** — the part of the crust that has not been in contact with the pan during baking.
**topfenstrudel** — *Germany* flaky pastry filled with creamed, vanilla-flavored white cheese, formed into a roll, and baked.
**top heat** — the heat a product receives from the oven parts above the product.
**topping applicators** — salters, sprinklers, depositers, and other devices that add (at a more-or-less constant rate) some decorative or flavoring material to the top of either a baked product or a shaped dough that is about to enter an oven.

**topping flour** — flour added at the dough stage, in the sponge-and-dough process for making bread.

**topping salt** — common salt (sodium chloride) that has been processed to be fairly uniform in particle size, and is usually in fairly large granules, used for depositing on the surfaces of soda crackers, pretzels, etc. iommediately before they enter the oven.

**torrada** — *Portugal* toast.

**torsion balance** — or "torsion scale," also known by various trade names. A weighing device that electronically measures the amount of twisting undergone by a metal bar, strip, or plate loaded with the object to be weighed; a variation of the spring balance in which the weight is balanced by the torsion of a wire.

**torta** — *Serbia/Croatia Portugal* swiss roll. *Italy* pie, tart, flan.

**torta de Viana** — *Portugal* swiss roll filled with lemon curd.

**törtchen** — *Germany* small tart or cake.

**torte** — *Germany* cake, usually round and often of the layer cake type.

**tortelettes** — *France* small round cakes.

**tortelli** — *Italy* small fritters.

**tortellini** — *Italy* ring-shaped pasta filled with meat, cheese, etc. Usually small in size.

**torten** — or tortes. Cakes of the continental type, especially large fancy cakes enriched with creams, marzipans, chocolate, fruits, and nuts; also used for pastries that resemble coffee cakes or pies.

**tortenguss** — *Germany* colored and flavored glazes (usually transparent) used as decorations and to prolong the shelf life of cakes and tarts.

**tortiglione** — *Italy* almond cake.

**tortilla** — *Spain* omelette or pancake. *Mexico* A thin, disc-shaped, unleavened flatbread; traditionally, hand-formed from lime-treated partly gelatinized corn kernels that have been ground to a paste, formed into thin patties, and surface baked (as on a griddle). There are now common versions made from wheat flour and often including small amounts of baking powder. oil, and other ingredients. Tortillas form the basis for many foods in Mexican cuisine, e.g., tacos, burritos, and enchiladas, but are little known in Spain and in some Central and South American countries.

**tortilla chips** — pieces cut or torn from tortillas and fried. Also, baked or fried extruded pieces of masa or corn meal resembling in one or more of their characteristics the traditional product.

**tortillina** — *Mexico* a word sometimes used to indicate tortillas made from wheat flour; for all practical purposes the word is unknown in the US.

**tortino** — *Italy* a kind of tart with savory fillings.

**tortita** — *Spain* waffle.

**tort makowy** — *Poland* poppy seed torte.

**tort piaskowy** — *Poland* sand cake.

**torttu** — *Finland* tort, flan, cake.

**tostada** — *Mexico* a tortilla fried crisp to provide a flat base for, e.g., mashed beans, lettuce, cheese, and cooked ground beef.

**tostato** — *Italy* toasted.

**total dietary fiber** — all of the various non-digestible vegetable components of a foodstuff, as determined by various tests, not all of them accurate or reliable.

**Total Fat** — according to 1993 labeling regulations, the sum of fatty acids from mono-, di-, and tri-glycerides, free fatty acids, phospholipid fatty acids, and sterol fatty acids expressed as triglycerides.

**totalizing flow meters** — fluid measuring devices that show the total amount of water (or other liquid ingredient from a bulk source) that has been added to, for example, a mixer bowl.

**total titratable acidity** — the total acid produced by fermentation, as determined by direct titration with standardized alkali solution.

**totopos** — *Mexico* tortilla chips; tortillas cut into pieces and fried crisp for dipping.

**Totox Value** — a determination of the degree of fat oxidation. The following formula is used: Totox = 2x(peroxide value) + anisidine value.

**tough** — a not-very-specific term used to describe the resistance of dough to stretching or tearing.

**tough wheat** — wheat grains having a moisture content between 14.0% and 15.5% for soft wheat and between 14.5% and 16.0% for hard wheat.

**tourte** — *France* torte.

**tower malt systems** — see "vertical malt houses."

**toxin** — a poison formed as a specific secretion product in the metabolism of a vegetable or animal organism; sometimes used to mean any substance that is deleterious to human health, but such usage is questionable.

**tragacanth** — a natural gum soluble in hot water that has been used as a component of glazes, as a binder in confectionery, and as a stabilizer for ice cream.

**tramp iron** — bits of metal found in loads of grain as it arrives at the mill.

**trans fatty acids** — signifies that, at the unsaturation points of a fatty acid molecule, the single hydrogen atoms on each of the carbons joined by double bonds are located on the opposite sides of the molecule. If on the same side, the term "cis" is applied.

**translucent** — not opaque but not entirely transparent; describes materials that transmit light with considerable internal dimming and distortion.

**transparent** — (1) Allowing the passage of radiation without substantially reducing its energy; in its widest sense applicable to both visible and invisible radiation. (2) The property of a material which allows observation of objects through intervening thicknesses of the material without substantial distortion or blurring, as in the case of sheets of clear glass.

**transpositor** — the mechanism in overhead proofers that inverts dough pieces about halfway through the proofing period.

**traveling hearth ovens** — ovens in which the hearth consists of steel plates linked together and mechanically drawn through the baking chamber. Straps of pans are placed on the hearth at one end of the chamber and the baked product removed at the other end.

**traveling tray and rack** — a type of baking plant that groups pans (or straps) and places them on trays, the latter being carried through the oven chamber by specially designed conveyors.

**traveling tray ovens** — ovens in which trays are pulled along a horizontal track by chains; the trays are loaded with panned dough pieces at the front end, travel from front to rear of the chamber, are transferred to a lower track, and are then carried back to the starting point (but at a lower level) where the baked product is unloaded.

**trays** — (1) Metal platforms attached to conveyors for carrying multiples of pans or straps through an oven. (2) Plastic containers into which bakery products are placed for transfer from the production point to sales outlets.

**treacle** — *UK* molasses.

**trenette** — *Italy* noodles.

**tricin** — a flavone; the principal coloring matter in wheat kernels.

**triethyl citrate** — a chemical that improves the whipping properties of egg products, under certain conditions.

**trifle** — a multiple-serving dessert traditionally made from a combination of cake, wine, macaroons, and fruit preserves. A typical modern example might consist of liqueur-soaked ladyfingers positioned around the inside of a glass bowl which is then filled with Bavarian cream; this assembly is sometimes topped with whipped cream and/or it may contain jelly or fruit preserves in a recognizable layer. Some trifle recipes are very elaborate.

**triglycerides** — chemical esters formed by combining one unit of glycerol with three units of fatty acids.

**trigo** — *Spain* wheat.

**trigo sarraceno** — *Spain* buckwheat.

**tristimulus** — a method of specifying color that relies on the quantification of three factors: e.g., hue, brilliance, and saturation. Other systems have also been called "tristimulus."

**triticale** — a plant that combines the genetic material of wheat and rye; the grain can be milled into a flour that makes fairly good bread.

**trolley cookies** — fairly soft base cakes that are dipped in successive coats of, e.g., marshmallow, icing, and chocolate couverture, by a device consisting of pins (to hold the cookies) mounted on an elaborate conveying device (the trolley).

**trough** — a large metal container of U-shaped cross section, usually on casters, for holding batches of dough during bulk fermentation and for transporting the dough between processing areas. Sometimes provided with simple mechanisms, or conveyors, for facilitating dough removal.

**trough elevator** — a trough hoist.

**trough hoist** — a motor-powered device for raising a dough trough above the mixer so the sponge can be easily transferred to the bowl.

**trypsin** — a protein-hydrolyzing enzyme found in animals' digestive tracts, and elsewhere.

**tsoureki** — Greek Easter bread, rich yeast-leavened dough formed into three cylinders then plaited; usually doesn't include fruits, but may include one hard-boiled egg in its (colored) shell in each large pastry.

**tsureki** — *Greece*bun.

**tulumbi** — *Serbia/Croatia* fritters served with syrup.

**tumma leipä** — *Finland* dark bread.

**tunnbröd** — *Sweden* unleavened barley bread.

**tunnel ovens** — ovens with baking hearths made of steel segments that move through the chamber; they are loaded at one end and discharge baked goods at the other end. Customarily used as one-product ovens in large wholesale bakeries.

**turbinado sugar** — raw sugar that has been partially refined by washing in a centrifuge to remove molasses from the surface of the granules. Darker and less purified than regular cane or beet sugar.

**turbidity** — in describing optical characteristics of a liquid, the extent to which light passing through the fluid is reduced in intensity by suspended materials such as clay, organic debris, and industrial wastes.

**turbine meters** — devices for measuring rate of fluid flow that depend upon the rotation of a vaned (like a fan) element as the sensing method.

**turbomilling** — converting grain into finely ground meal by impact methods and then separating the meal into fractions having varying properties (e.g., high- and low-protein content) by air-classification equipment.

**türkenkorn** — *Germany* maize.

**turmeric** — a spice made from the dried root of the plant *Curcuma longa*. Used mostly for its potent yellow coloring effect, although it does also contribute a flavor somewhat resembling a combination of ginger and pepper.

**turn** — in the production of puff pastry dough, the operation of folding the dough in layers after each sheeting operation, usually after turning the piece 90 degrees from the preceding sheeting operation.

**turnover** — (1) The rate at which fat is used up in a frying operation. (2) A single-serving dessert product made of a disc or square of puff pastry dough, one-half of which is covered with a fruit (or other) filling, the uncovered half being folded over to conceal the filling, and the ensemble then baked.

**turntable** — a rotatable cake stand used by decorators to simplify application of icing and the like to a cake.

**túrós töltelék** — *Hungary* cottage cheese filling, as for pancakes.
**tutti frutti** — *Italy* mixed glazed fruits.
**Tweedy mixer** — a dough mixer of special design used for very high intensity development of bread doughs and the like, provided with means for drawing a partial vacuum during part of the mixing process.
**twin rolls** — bread rolls made by placing two dough balls of the same size side by side in a muffin-cup type of pan, then proofing or baking; also, rolls formed by any other means that result in the same finished shape of two equal lobes.
**twist** — designates a type of weave used in fabrics for sieves. In half-twist cloth, every alternate warp consists of two half-size threads, one passing over the woof and the other under it. In full-twist, every warp consists of two half-size threads.
**twist bread** — a method of molding bread that consists of forming two cylinders, each half the desired weight of the panned loaf, then twisting them around each other before they are placed in the pan; believed to improve the appearance and texture of the finished loaf.
**twisted goods** — spaghetti or vermicelli formed into twisted clumps before drying, the advantage being that long strands can be packaged in compact containers with minimal breakage.
**twister** — a twist molder, or the parts of that machine that perform the actual twisting after the dough cylinders have been formed by other parts of the molder.
**twist molder** — a production machine that forms cylinders of bread dough, twists two of them together to form a loaf-size piece, and places the loaf in a baking pan.
**two-row barley** — a common type of barley in which only one row of kernels develop on either side of the plant spike, contrasted to six-row barley.
**tyesto** — *Russia* dough or batter.
**tyesto byezdrozhzhevoye** — *Russia* unleavened dough.
**tyesto drozhzhevoye** — *Russia* leavened dough or batter.
**tyesto sloyonoye** — *Russia* puff pastry dough.

### -U-

**udon** — *Japan* noodles made of wheat flour and water; also, the edible shoots of a certain Japanese herb.

**ugnsbakad** — *Sweden* baked.

**ugnspannkaka** — *Sweden* a kind of batter pudding.

**ukoy** — *Philippines* a fritter of ground rice or flour, diced scallions, and small peeled shrimp.

**ultrafiltration** — a process sometimes applied in the treatment of water to remove particulate matter of very small size by forcing the water through semi-permeable membranes.

**ultraviolet light** — the short wave lengths of electromagnetic radiation occuring just beyond the blue or violet end of the visible spectrum. UV radiation has some sterilizing capabilities.

**umbrella rounder** — a type of rounder for dough pieces that has a rotating cone with its apex at the top.

**unbolted** — not sifted, applied to wheat meal and the like.

**uncertified color additive** — a color additive approved for food use but not listed as a certified color. Examples are annatto extract, titanium dioxide, and beet powder.

**unctuous** — greasy, as an ointment; having the most notable physical characteristics of oil and fat, hence, rich in lipids or appearing to be so.

**uniformity coefficient** — as a sieve specification, the value obtained by dividing the width of the orifice that will pass 60% of the sample by the width of the opening that will pass no more than 10%.

**unit operation** — a chemical engineering term identifying a physical change to which material is subjected especially in connection with a unit process; e.g., the rounding of dough pieces in a bread loaf preparation line.

**unit process** — a chemical engineering term identifying a chemical change to which material is subjected as a step in a manufacturing procedure; the fermentation step in breadmaking is an example.

**United States Standards** — for grain, those specifications that must be followed when describing shipments of grain entering interstate commerce.

**unleavened** — a dough or batter that does not contain an ingredient added for the purpose of generating gas within the mixture; it is the usual practice to describe doughs or batters that do not contain yeast, baking powder, ammonium bicarbonate, etc., as "unleavened" even though they may expand considerably in the oven as the result of the increased internal pressure of water vapor and air.

**unloaders** — equipment that removes pans with their finished baked products from the oven and transfers them to conveyors leading to subsequent operations, such as depanning.

**unsaponifiable matter** — substances found in natural fats and oils that cannot be saponified, i.e., do not split when reacted with aqueous alkalies. Includes some of the more complex organic chemical compounds that are commonly soluble in ordinary fat solvents.

**unsaturated** — (1) A liquid not containing the maximum amount of a substance that could be dissolved in it under the conditions of observation. (2) A term descriptive of the distribution of hydrogen atoms on the carbon "backbone" of a fatty acid (chemically, the term has a much broader meaning). In a fully saturated molecule, each of the carbon atoms (except the end atoms) have two hydrogen atoms attached. If the molecule is mono-unsaturated, two adjacent carbon atoms in the molecule will each have only one hydrogen atom attached. In polyunsaturated fats, more than one pair of adjacent carbon atoms will lack the full complement of hydrogen atoms.

**unsavory** — unpleasant in taste or odor.

**unsulfured molasses** — molasses that has not been treated with sulfur dioxide during its processing; usually applied to the so-called whole cane juice molasses, a premium grade.

**uovo** — *Italy* egg.

**upside-down cake** — a cake, usually square in shape, made of a thick, fairly rich chemically leavened batter baked on top of a mixture of (typically) canned crushed pineapple and brown sugar. Served inverted, so the pineapple layer is on top. Other fruits can be used.

**uunissa paistettu** — *Finland* baked..

**uva pasa** — *Spain* raisin.

## -V-

**vacherin** — *France* a meringue shell filled with, e.g., ice cream.
**vacuum cooler** — a chamber in which hot bread loaves are placed, there being subjected to a partial vacuum so that the loaves cool by the evaporation of some of the water they contain.
**vacuum depanners** — machines for removing baked loaves, buns, and rolls from their pans by contacting them with a vacuumized nozzle that is then moved so as to deposit the baked piece in a desired location.
**vaffel** — *Denmark* either a waffle or a wafer cookie.
**vafle** — *Serbia / Croatia* waffle.
**vaflya** — *Russia* waffle.
**vainilla** — *Mexico* vanilla.
**valkoinen leipä** — *Finland* white bread.
**välling** — *Sweden* soup containing considerable cereal meal; porridge.
**valmuefrokage** — *Denmark* poppy seed cake.
**valnöt** — *Sweden* walnut.
**vand** — *Denmark* water.
**vaniglia** — *Italy* vanilla.
**vanilj** — *Sweden*
**vanilla** — the flavor prepared from cured vanilla beans, usually by extracting them with ethanol solutions. Imitation vanilla flavor consists of vanillin with other synthetic and natural materials such as ethyl vanillin.
**vanilla beans** — the fruit of a tropical orchid. The long pods are put through a "curing" process, during which enzymic and other changes lead to development of the typical vanilla flavor.
**vanilla extracts** — solutions of various concentrations (single-fold, double-fold, etc.) of the alcohol soluble materials removed from vanilla beans; of course, there are also "imitation" vanilla extracts.
**vanille** — *Germany Netherlands* vanilla.
**vanillin** — a chemical substance, 4-hydroxy-3-methoxy-benzaldehyde, responsible for the top note of both natural and artificial vanilla aroma.
**vanukas** — *Finland* pudding.
**varietal improvement** — enhancing the commercially important characteristics (such as yield per acre or flavor) of varieties of (e.g.) grain crops by breeding programs, hybridizing, selection or the like.
**varm** — *Denmark* warm.
**varyeniki** — also, "varenyky." *Russia* dumplings or ravioli made from a flour, cottage cheese, and egg yolk dough, with fillings such as sauerkraut, berries, chopped meat, or cheese. Cooked by poaching or boiling.
**vasilopita** — *Greece* cake flavored with orange or mastic (an aromatic resin) and traditionally served on New Year's Eve.

**vatruchka** — *Russia* small tarts of cottage cheese.

**vatten** — *Sweden* water.

**vegetable colors** — materials obtained from plants, used for coloring foods; beet powder is a typical example. Commercial vegetable colors have all been purified and concentrated to a greater or lesser degree.

**vegetable muffins** — muffins of the cupcake form based, usually, on a moderately sweet chemically leavened batter containing bits and shreds of various vegetables or, sometimes, vegetable juices.

**vegetable parchment** — a tough sheet made by immersing uncoated paper in sulfuric acid, then washing and drying the sheet. Grease resistant and expensive; responds poorly to most sealing methods unless it is coated.

**vekna** — *Serbia/Croatia* loaf.

**velli** — *Finland* gruel, porridge.

**velocity** — (1) Speed (unit of length per unit of time) at which a solid or fluid is moving past a given point; rate of motion; distance moved in unit time. (2) Rate at which a chemical reaction progresses.

**ventilate** — to increase the movement of ambient air through a space, e.g., by making openings in an enclosure to facilitate natural air currents or by using fans.

**venturi meters** — devices for measuring the velocity of fluid moving in pipes.

**vermicelli** — *Italy* pasta of the spaghetti type, very thin in cross-section, often sold as "nests," i.e., many strands wound together to form a bundle.

**vertical malt houses** — malt production plants in which barley enters the top of a tower and undergoes the steps of steeping and sprouting as it gradually travels toward the bottom of the bin.

**vertical mixer** — any mixer having an agitator suspended in a vertical position; planetary mixers are all vertical mixers, but not all vertical mixers are planetary mixers.

**vesi** — *Finland* water.

**vesicles** — a fairly common term used to describe the bubbles or cells forming the internal structure ("crumb") of bread, etc.

**vett** — *Estonia* water.

**vibrating hoppers** — arrangements for facilitating the movement of a powdered or granular ingredient from a bin or silo to a delivery point, usually consisting of a cone-shaped chute with an electric motor attached.

**vibrating screw feeders** — a transporting mechanism consisting of an auger-type conveyor rotating in a tube at a variable rate, some or all of the parts of which are vibrated to reduce packing and flooding of the conveyed material.

**Victoria sandwich** — *UK* a type of layer cake filled with preserves.

**Vienna bread** — a hearth-type yeast-leavened bread in loaf form with heavy crisp crust, made from a lean dough. The loaf is usually elongated and somewhat pointed at the ends, and the top may be slashed (usually lengthwise.)

**Viennoisierie** — *France* Vienna-type bakery products.

**viipale** — *Finland* slice.

**vijg** — *Netherlands* fig.

**vinegar** — a dilute (usually 5%) solution of acetic acid produced either by microbiolgical action on fermentation alcohol to give cider, malt, or wine vinegar, or by chemically oxidizing essentially pure ethanol from synthesis operations to give white or distilled vinegar.

**vinagre** — *Spain* vinegar.

**violets** — candied violet flowers were formerly used as decorations for fancy baked products, but have been replaced by artificial blossoms made from colored sugar icing.

**viscometer** — any of a number of instruments designed to measure the viscosity or apparent viscosity of fluids.

**viscosity** — a liquid's resistance to flow.

**viscous** — having a relatively high resistance to flow.

**vispgrädde** — *Sweden* whipped cream.

**vital wheat gluten** — purified gluten prepared from wheat in such a manner that many of the native properties (including extensibility) are retained; contrasted to denatured wheat gluten, which will not form a cohesive extensible mass when rehydrated.

**vitamin premixes** — proprietary materials containing sufficient vitamins to adjust a given batch size of a given product to a legally-defined level of nutrient content, often with sufficient overages to account for loss during processing or storage.

**vitamins** — organic substances required in trace amounts for catalyzing or otherwise supporting normal metabolic processes in living organisms and, usually, not synthesized by the organisms in amounts sufficient to promote optimum health. Different animal species may require different vitamins in their diet.

**vitlök** — *Sweden* garlic.

**vitreous** — glassy. As one example, used to describe the cut surface of a kernel of hard wheat, when it refers to a somewhat shiny, slick, translucent appearance.

**vlaai** — *Netherlands* fruit tart.

**voda** — *Russia* water.

**vohveli** — *Finland* wafer, waffle.

**voi** — *Finland* butter.

**voileipä kekksi** — *Finland* soda cracker.

**vol** — *UK* a commercial mixture of ammonium carbonate, ammonium carbonate, and ammonium carbamate; it is used primarily as a leavener in cookies.

**vol au vent** — a circular puff pastry shell with its center open at the top, like a cup, sometimes with a lid of puff pastry. The pre-baked shells are filled either with savory preparations such as creamed chicken or with fruit and whipped cream.

**volumetric measuring** — any kind of metering process that is based on the determination of the space occupied by a known weight of material, thus inextricably connected with, and requiring, uniform density of the ingredient.

**votator** — a device for plasticizing fats; basically a scraped surface heat exchanger through which the melted fats are pumped. Heat is removed from the fat until crystallization or solidification of some of the fractions occurs. The solidified or plastic material exiting the votator is of homogeneous composition and of a consistency permitting it to be molded or packaged into cartons as though it were a solid product. The material also gives different results (usually) than the liquid oil constituents when it is used as an ingredient in cakes and cookies.

**vrucht** — *Netherlands* fruit.

**vutiro** — *Greece* butter.

**vyermishel** — *Russia* thin noodles, vermicelli.

## -W-

**wafel** — *Netherlands* wafer
**wafer** — (1) A thin cookie. (2) The crsip baked sheet used as a component in sugar wafers. (3) A thin tablet or pellet containing a precise amount of nutrient additives or dough modifiers, often used for supplementing bread doughs.
**wafer ovens** — automatic ovens that manufacture the wafer sheets used to make sugar wafer cookies.
**wafer sheets** — the large thin sheets baked from a thin batter consisting mostly of flour that are sandwiched with sugar-fat creams and then cut into individual sugar wafer cookies.
**waffeln** — *Germany* sugar wafers.
**waffle** — a firm, fairly thick cake (soft inside, crisp outside) with deep, square surface indentations, baked in a special mold. Somewhat similar to a pancake in composition and usage although it is considerably lower in moisture content than a regular pancake and the batter frequently contains some eggs and milk.
**walking finger conveyors** — a type of conveyor used mostly for unloading pies and other delicate products from ovens; the functional principle is the sequential movement of two sets of bars set perpendicular to product travel and on which the products rest.
**walnoot** — *Netherlands* walnut.
**walnut** — (1) The English walnut, *Juglans regia*. (2) The black walnut, *Juglans nigra*.
**warm** — *Netherlands* hot.
**wash** — a liquid brushed on the surface of a baked or unbaked product to alter its crust characteristics (appearance, texture, flavor, etc.).
**washover** — *UK* a wash.
**wasser** — *Germany* water.
**water** — chemically, $H_2O$; practically, a component found in every food and an ingredient found in nearly all foods. It is an important determinant of texture, appearance, flavor, nutritional value, and storage stability of foods and beverages. Unless you know the quantity and characteristics of your water supply, you cannot have a complete knowledge of the processing responses and finished quality of your food product.
**water absorption** — the percentage of water, relative to flour as 100%, required to yield a dough of the desired consistency.
**water activity** — ratio of the relative humidity of the atmosphere in equilibrium with a water-binding substance as compared to the relative humidity of the atmosphere above pure water held at the same temperature and pressure.

**water brews** — fermentation mixtures or "broths" containing no flour.

**water icings** — generally simple mixtures of water, powdered sugar, colors, flavors, and gums (no fatty materials) used to coat the tops of cookies and the like.

**Water Quality Act** — a federal law passed in 1967 that required every state to establish quality criteria for all interstate waters.

**water softening** — any method that reduces, in a water supply, the mineral ions (principally calcium ions) responsible for reacting with soap constituents to decrease the soap's effectiveness and to form scum.

**water source** — the originating point of a water supply system, e.g., lakes, rivers, reservoirs, and wells.

**water splitter** — a device that uses a high velocity jet of water to cut a slit in the top of bread dough that is ready for the oven.

**water vapor leavening** — the aeration or "raising" of a dough or batter as the result of internal pressure due to heat-induced increase in the vapor pressure of water.

**wax** — a relatively tough, meltable, solid material insoluble in water; a typical example is beeswax. Synthetic and natural kinds are available. Waxes are generally not digestible.

**waxed paper** — thin paper coated on one or both sides with petroleum wax, sometimes supplemented with plastic resins. Once widely used for packaging bread loaves, but now rarely so used.

**waxy** — an adjective that, when applied to grain, indicates there is a greater proportion of amylopectin in the endosperm than in the endosperm of ordinary varieties of the grain.

**waxy maize** — a variety of corn having nearly all of its starch present in the form of branched molecules (amylopectin).

**weissbrot** — *Germany* white bread.

**weizen** — *Germany* wheat.

**weizenbrötchen** — *Germany* bread rolls of various types, yeast-leavened.

**weizenmischbrot** — *Germany* the German standard bread consisting of a mixture of more than 50% wheat flour with a lesser amount of rye flour. Given a relatively short fermentation.

**welschkorn** — *Germany* corn (maize).

**Welsh rabbit** — melted cheese (or cheese blended with spices and other flavors and condiments) on toasted bread. Many variations and elaborations are known.

**wentelteefje** — *Netherlands* French toast.

**Westphalian pumpernickel** — the darkest, densest, dankest kind of pumpernickel; seldom made in the U.S.A.

**wet milling** — a process for separating corn kernels into their component parts using a water-sulfur dioxide treatment in combination with milling and separation equipment.

## GLOSSARY OF CEREAL TECHNOLOGY TERMS    213

**wet peak** — a stage in the egg whipping process; when the mixing utensil is withdrawn from the meringue (for example) the peak of the mixture will glisten with free liquid and will, usually, soon fall over.

**W-folding** — a type of cross-panning of bread that involves bending the cylinder of dough to form a "W," said to improve uniformity of crumb.

**wheat** — the plant of the genus *Triticum*, and its seed. Many species and variants are known, such as common bread wheat, durum, etc.

**wheat bread** — bread made from a dough containing both whole wheat flour and white flour.

**wheat class** — one of the accepted commercial categories of wheat kernels, e.g., soft red, hard red, white, etc.

**wheat flakes** — a processed (ready-to-eat) breakfast cereal, usually made by moistening and tempering whole kernels, flattening the grain slightly, cooking with sugars, salts, and other ingredients, flaking, then oven-drying and toasting.

**wheat germ** — the germ or embryo of the wheat seed; contains a fairly large amount of oil and oil-soluble vitamins. Also contains substances that weaken gluten.

**whey** — the residue of milk after the butterfat and curds have been removed in the preparation of cheese; consists principally of water, albumin proteins, inorganic salts, and lactose.

**whey protein concentrates** — commercial preparations made by drying whey from which most of the minerals and lactose have been removed.

**whey protein test** — usually means a test of the extent to which the whey proteins have been denatured during drying, often correlated with the baking performance of doughs containing the whey preparation.

**whipped cream** — cream of high butterfat content that has been whipped or beaten to make it incorporate large amounts of air; similar product dispensed from a pressurized container. Whipped cream usually contains sugar and vanilla flavor and, often, stabilizers.

**whipping** — beating a fluid or semi-fluid material such as egg whites or cream to incorporate air and then to subdivide the bubbles into a fine and uniform foam.

**whipping aids** — any of several chemical substances that will, in certain circumstances, improve the rate and extent of aeration of a foamed ingredient such as egg white.

**white layer cakes** — a term usually reserved for chemically leavened layer cakes that contain a significant amount of egg white and no egg yolk or whole eggs.

**whitening cones** — machines used in rice mills for removing the brown layers from rice by the abrasive action of cone-shaped rotating elements.

**white pepper** — a ground preparation made from decorticated black pepper seeds.

**white wheat** — a soft wheat with a light-colored hull and a relatively friable endosperm; suitable for pastry flours. Grown in relatively small quantities in the US.

**whizzing** — whirling grain in a special centrifuge to rid it of surplus water that has been added during a washing process.

**wholemeal** — *UK* wholewheat.

**whole wheat bread** — bread made from dough that has its entire wheat content as whole grain meal or flour.

**wicket** — a simple metal attachment for holding the stack of plastic bags used in automatic bread bagging machines.

**wicket hole** — the small holes punched in a plastic bread bag that allow it to be threaded on to the wicket.

**wicket pack** — the wicket with its complement of 100 or 200 plastic bread bags, as delivered by the supplier.

**wiener** — (1) A baked product allegedly of Viennese or Austrian type or origin. (2) A sausage of the hot dog shape.

**wienerleipä** — *Finland* Danish pastry.

**wienerbrod** — *Denmark* Danish pastries. Also, *Sweden* "wienerbröd."

**wienerbrodsdejg** — *Denmark* dough for Danish pastry (or, as they call it, Vienna bread).

**wild break** — unusually large break and shred at the side of the baked loaf, often irregular. This fault is not only unsightly, but it suggests the likelihood that other defects will be found in the bread's texture and uniformity.

**wild rice** — a grain that is botanically quite different from ordinary rice, and is also different in appearance, flavor, and texture. Has a reputation as a gourmet replacement for ordinary rice. Botanically, *Zizania aquatica*.

**wild yeast** — basically, any yeast other than *Saccharomyces cerevisiae* that unexpectedly contributes to leavening and flavor development in a dough. The changes are nearly always for the worse.

**Wiley melting point** — a specialized test for fats that measures the liquefaction of a disc of fat heated under certain conditions.

**windbeutel** — *Germany* cream puff.

**windmill cookies** — molded cookies in the shape of a windmill, usually they are brown in color, highly spiced, and contain sliced almonds.

**wine** — fermented fruit juice; commercial versions often contain additives that are not declared on the container. If made from any fruit except grapes, a modifier must be added to the name, e.g., peach wine, cherry wine.

**wine ball** — also, "wine cube." *China* Small balls consisting mostly of compressed brewers' yeast; sold in markets for consumer use in making fermented beverages.

**winnow** — to separate and drive off chaff by means of wind; to fan, as to winnow grain.

**winterization** — a process applied to food oils to remove naturally occurring high-melting triglycerides so that the oil will not become cloudy during cold storage. The oil is held at a reduced temperature until the crystallization of high melting point fats is complete, and then the solidified fats are filtered out.

**winter wheat** — wheat that is sown in the late fall or early winter and harvested in the late spring or early summer; most U.S.A. wheat is of this type.

**wire-cut cookies** — the dough for these cookies is extruded as a continuous cylinder through an orifice; the dough strand is cut cross-wise by a wire (or band) drawn across the orifice at regular intervals. The slices of dough so formed fall directly on the oven band. Various cross-sectional shapes are possible.

**WONF** — "with other natural flavors," descriptive of compounded flavoring materials that contain both the characterizing natural material (e.g., strawberry juice for strawberry flavors) and some other natural substances that improve or intensify the desired flavor (e.g., raspberry juice, orange oil, citric acid).

**wontons** — also, "wan tons," "Chinese ravioli," and "cloud swallows." *China* A thin, unleavened dough sheet folded around a relatively small amount of filling such as shellfish meat, then cooked in various ways, such as by frying, boiling in soup, etc.

**wonton sheets** — thin squares, about 4x4 inches, of unleavened dough (characteristically now made from wheat flour though originally perhaps from rice flour), used to fold around meat or other fillings to make wontons, but also occasionally fried separately to form a kind of snack or appetizer.

**wort** — the liquid mixture prepared for fermentation in the brewing process or as an intermediate product in the manufacture of malt syrup. Consists largely of water, carbohydrates (mostly fermentable sugars), proteins, and other materials extracted from malt or added to the brewing kettle as ingredients.

**WURLD process** — a method of lye-peeling of cereal grains intended to substitute for the more elaborate bran-removal processes used in standard mills. Has been applied mostly to rice.

**WVTR** — water vapor transfer rate, an important specification for packaging materials, especially for plastic films.

## -X-

**xanthan gum** — a gum produced by the action of the microorganism *Xanthomonas campestris* on glucose. When combined with locust bean gum, a synergistic reaction occurs that leads to gel formation.

**xanthein** — a soluble coloring matter found in some yellow flowers, in the past used or suggested as a possible color for cereal foods.

**xanthophylls** — chemically distinct kinds of pigments found in many plants, including wheat. Vary in color from light yellow to dark brown.

**xoi nep** — *Vietnam* cooked glutinous rice of a type often called "sweet rice."

**x-ray crystallography** — a method for determining the physical arrangement of atoms or molecules in crystalline materials. It can be used for studying the polymorphic form of a solidified fat, for example.

**xylan** — a yellow gummy pentosan, found in some woody plants.

**xylanase** — an enzyme that splits xylans into their component pentoses; it has been suggested as a dough-improving additive.

**xylitol** — a pentahydroxalcohol or sugar alcohol, formed by hydrogenating sugars; it has a mildly sweet taste and is not fermented by bakers' yeast.

**xylose** — a non-fermentable, five-carbon sugar. D-xylose is formed by the hydrolysis of xanthan, and has been called "wood sugar."

## -Y-

**yadan** — *China* egg.

**yaki fu** — *Japan* larger sizes of nama-fu.

**yaourtopita** — *Greece* a kind of cake made from a dough containing mainly flour and yoghurt.

**yaytsa** — *Russia* eggs.

**yeast** — the unicellular microorganism that is responsible for leavening bread, etc. Also, commercial preparations consisting mostly of these cells.

**yeast food** — a mixture designed to be added to a dough to increase yeast activity and, particularly, to improve cell multiplication. Many of these proprietary products consist mainly of mineral salts.

**yeast-leavened products** — those bakery foods for which the dough or batter has attained a necessary part of its aeration (increase in volume) through the carbon dioxide released as a result of the fermentation of sugars by yeast.

**yeast substrate** — a substance used by yeast for its metabolic processes, including those, like sugar, transformed by fermentation into alcohol and carbon dioxide.

**yellow cakes** — chemically leavened layer cakes, the crumb of which has a yellowish hue, this being, in earlier times, an indication of the use of significant amount of whole eggs or egg yolks in the batter. Traditionally flavored with vanilla and a small amount of lemon extract.

**yellow popcorn** — the most common type of popcorn, the exterior layers being colored yellow, but the endosperm being essentially white, like all other varieties of popcorn.

**yema** — *Spain* egg yolk.

**yield** — in general, the amount of any product or intermediate resulting from the processing of a given amount of ingredients or intermediate products, expressed either as a percentage of the beginning materials or as final weight. Specific examples are: the amount of finished mill products expressed as a percentage of the amount of grain required to make them, and the weight of baked product resulting from 100 pounds of ingredients.

**yit bien** — *Chi* moon cake, a baked bun with filling; chemically leavened.

**yogurt** — also, "yoghurt,""yoghourt," "yaourt," etc. Milk thickened by acids that have been developed in the fluid through bacterial action. Modern consumer types are nearly always sweetened and flavored with additives. Yogurt has been used as an ingredient in doughs, batters, and fillings.

**yolk** — the yellow part of a chicken's egg. Available commercially in many forms, e.g., pasteurized, frozen, dried, and with and without adjuncts. A valuable source of high quality protein; also contains emulsifying ingredients, fat, and cholesterol.

**young** — describes a yeast dough that is underfermented. These doughs produce baked products that are tight in grain, low in specific volume, and relatively tough.

**you tiau crullers** — *China* deep-fried pastries, like doughnuts but in long shapes, that are sliced and served with congee and the like.

**yufka** — *Turkey* this name has evidently been applied to more than one kind of product, but a common form is a thin, round flatbread made from an unleavened dough.

-Z-

**zakuska** — *Serbia/Croatia* snack.

**Zante currant** — a kind of seedless grape dried to make a small raisin.

**zarzamora** — *Spain* blackberry.

**Zea mays** — the genus and species names of corn or maize, including many varieties having an extraordinary range of culinary properties.

**zein** — the predominant protein (group) of corn, a prolamine.

**Zeleny sedimentation test** — a wheat quality test based on the amount of sediment that forms in a graduated cylinder in which has been placed a flour suspension prepared under carefully controlled conditions.

**zeljanica** — *Serbia/Croatia* tart made of layers of pastry and spinach.

**zèppola** — or, zeppoli. *Italy* a rich fritter, doughnut, or cruller, often filled or topped with, e.g., custard. A St. Joseph's Day specialty.

**zest** — the outer portion of fresh orange or lemon peel, used in strips or finely grated as an ingredient for its aromatic oil content.
**zhir** — *Russia* shortening, fat.
**zimicka** — *Serbia/Croatia* roll, bun.
**zimt** — *Germany* cinnamon.
**zito** — *Serbia/Croatia* wheat.
**zitrone** — *Germany* lemon.
**zone** — divisions or modules into which an oven may be separated along its length so as to give differing heat (or humidity) treatments to the dough pieces as they pass through the oven.
**zone fryers** — in long conveyor fryers (as for large-scale production of doughnuts), the separation into regions along the fryer's length, the different sections containing fat held at different temperatures.
**zout** — *Netherlands* salt.
**zucchero** — *Italy* sugar.
**zucker** — *Germany* sugar.
**zumance** — *Serbia/Croatia* egg yolk.
**zumo** — *Spain* juice.
**zuppa inglese** — *Italy* sponge cake containing candied fruit and steeped in rum, then combined with custard or whipped cream. Other variations are known, including some without the cake.
**zweiback** — see "zwieback."
**zwieback** — this word is so often misspelled "zweiback" that the latter version should perhaps be regarded as having developed into an acceptable alternative spelling (at least in the US). Zwieback is a kind of rich bread or coffee cake that is first baked in loaf form (often with a circular cross-section) and then sliced and toasted to a low moisture content under conditions allowing only slight to moderate browning. Typically, the texture is crisp, the color light brown, and the flavor fairly bland.
**Zygosaccharomyces** — a genus of yeast, in the same family as common bakers' yeast, that has been investigated as a possible replacement for bakers' yeast.
**zymase** — a combination of enzymes (as from yeast) that, acting in concert, transform certain sugars into carbon dioxide and ethanol.

www.ingramcontent.com/pod-product-compliance
Lightning Source LLC
Chambersburg PA
CBHW021140230426
43667CB00005B/202